Synthetic Biology

Synthetic Biology

Science, Business, and Policy

Lewis D. Solomon

Routledge
Taylor & Francis Group

LONDON AND NEW YORK

First published 2012 by Transaction Publishers

2 Park Square, Milton Park, Abingdon, Oxfordshire OX14 4RN
711 Third Avenue, New York, NY 10017

Routledge is an imprint of the Taylor & Francis Group, an informa business

First issued in paperback 2017

Library of Congress Catalog Number: 2011014821

Library of Congress Cataloging-in-Publication Data

Solomon, Lewis D.
   Synthetic biology : science, business, and policy / Lewis D. Solomon.
      p. cm.
   Includes index.
   ISBN 978-1-4128-1856-8
 1. Cells—Simulation methods. 2. Artificial cells. I. Title.
   QH585.S54 2011
   571.6—dc23
                        2011014821

ISBN 13: 978-1-4128-1856-8 (hbk)
ISBN 13: 978-1-138-51517-8 (pbk)

In memory of
Jack N. Bonné, my friend
Harold P. Green, my friend and colleague
Arthur S. Miller, my friend and colleague

Contents

1

Introduction

We stand at the threshold of a momentous scientific event, the writing of life, using the tools of synthetic biology. Today, researchers build on two of modern science's greatest developments: first, the 1953 discovery of the structure of DNA, the genetic text that spells out the recipe for life, and second, the reading of life, culminating in the mapping of the human genome, our entire genetic makeup, some fifty years later. Scientists now seek to write life, ushering in a wide range of applications, including groundbreaking changes in medicine, commerce, and daily life.

Synthetic biology dramatically transforms existing genetic modification techniques. For nearly forty years, using recombinant DNA (rDNA) tools, first researchers and then businesses have genetically engineered organisms. Recombinant DNA changes a living thing's genome in various ways by manipulating and transferring one or several naturally occurring genes, long stretches of DNA that specify how to make proteins, the building blocks of life, from one organism into another, thereby modifying the components of living cells to impart new traits and achieve desired functions, including the production of proteins, pharmaceuticals, and seeds.

Synthetic biology permits more complex and sophisticated engineering than can be achieved through previous genetic modification techniques. Drawing on nonbiological scientific and engineering disciplines, including information technology and nanotechnology, synthetic biology strives to rearrange an organism's genes on a far wider scale—its genome—by rewriting its genetic code, all the chemical instructions need to design, assemble, and operate a species. By manipulating whole living systems, synthetic biotechnology transforms existing organisms for more useful purposes and eventually will create new ones. By constructing new metabolic—energy production—pathways, researchers seek to create platform microorganisms capable of carrying out various functions and practical tasks, making things we

1

want, including new fuels, petrochemicals, and pharmaceuticals. A current aspiration of scientists focuses on altering microorganisms' genomes to produce living machines designed to turn algae or sugarcane into biofuels and chemicals. For-profit businesses look to scale-up and commercialize these genome-driven metabolic engineering discoveries, while obtaining funding to bring good ideas to fruition.

Ultimately, researchers hope to design and build from scratch novel, functioning biological systems, free-living life forms with life-like cells. This branch of synthetic biology aims to engineer living entities that do not exist in nature. In the future, we will witness the creation of completely artificial organisms, unconstrained by any existing template. Devising new life forms via biochemistry, engineered on the computer and made from off-the-shelf chemicals, will not only revolutionize biology, it will also profoundly influence the definition of life, including what it means to be human.

Exemplifying these two aspects of synthetic biology, namely, transforming existing organisms and creating new biological entities, this book focuses on the efforts of J. Craig Venter, Ph.D., one of modern biology's most innovative scientists and one of the most polarizing. In 2009, Venter received the National Medal of Science,[1] the highest honor awarded to a scientist by the U.S. Government. President Barack Obama presented the medal to Dr. Venter in recognition of his outstanding contributions to science, his dedication to genomics, the study of living things in terms of their genomes, his contributions to understanding the societal implications of genomics, and his commitment to clearly communicate information to the scientific community, the public, and policymakers.

Venter made his reputation in sequencing the human genome, the complete genetic information that controls how our bodies function. The term "sequencing" connotes the deciphering the genome's combination of string of DNA building blocks that chemically encode the core information directing the assembly and operation of every living thing. Announced in a blaze of publicity in June 2000, human genome mapping resulted from the work of two groups, Celera Genomics Corp., a for-profit company Venter then headed, and the mostly publicly funded Human Genome Project that used a team of scientists from around the world.

After successful decoding of the human genome, today, Venter, a prominent scientist-entrepreneur, remains as ambitious as ever. He seeks to find new ways to produce biofuels, by turning algae into petroleum, thereby revolutionizing the energy and petrochemical industries. He also wants to make a microbe from scratch that will herald the creation of new life forms in the laboratory. As he put it, "We're [his research institute and for-profit corporation] trying to solving the energy crisis and create life."[2] However, these minimal genome microorganisms, alive through human invention, fuel fears of environmental contamination and the development of biological weapons.

Beyond the scope of this book is an analysis of another branch of synthetic biology, BioBricks, a registry of standardized biological parts, DNA sequences, and other molecules that encode basic biological functions. In brief, these interchangeable parts (and other standard genetic units), researchers hope, will be pieced together into a variety of large functioning devices capable of performing bio-engineered tasks.

Overview of the Book

Chapter 2 examines three explanations of origins of earthly life, namely, chemical self-assembly, panspermia (meaning seeds everywhere), and superbugs. The superbugs theory looks to life's origins in hot, deep ocean locations, with the subsequent direction upward to the earth's surface. Recent discoveries, particularly Venter's mapping the genome of a sea-floor microbe, not only proved the existence of a third branch of species, but also supported the superbugs concept of the beginnings of earthly life.

Chapter 3 provides an overview of the chemical basis for life as revealed by molecular biology. The chapter focuses on cells—the basic biological unit of life, the chemical composition and structure of DNA—the instructions on how to make each species—and its structure, and how organisms manufacture proteins. Biotechnology applications are briefly considered.

Chapter 4 focuses on the steps involved in reading the genetic code. The quest to understand how all living things are assembled and operate achieved fruition, in part, through many of Venter's efforts from 1982 to 2001, cumulating in mapping the human genome. The chapter examines Venter and his team's successes first at the National Institutes of Health (NIH), then at The Institute for Genomic Research (TIGR), largely funded by a for-profit firm, Human Genome Sciences, Inc. (HGS), and finally at another for-profit venture, Celera Genomics Corp. (Celera). The chapter provides an overview of Venter's various organizational structures, his research team's scientific achievements, and the reasons his two post-NIH arrangements ended. The chapter briefly concludes with a discussion of the practical applications of genomic sequencing, particularly in medical research, diagnosis, and treatment.

Chapters 5 through 8 examine the quest to write life, including retooling the metabolic pathways in organisms and the creation of artificial life, forged by human intelligence. Chapter 5 discusses Venter's post-Celera organizational structure and the funding of his nonprofit entities. Chapter 6 provides an overview of the research progress in synthetic biology made by Venter's nonprofit team, culminating in the creation of the first synthetic cell. Two areas are considered: basic genomic and environmental genomic research. Researchers today design, synthesize, and assemble genes and entire genomes from the bare chemical components of DNA. In the future, scientists, led by Venter's team, will design an organism on a computer, beginning with the minimal genome, the smallest set of genes required for life, make the necessary DNA, and put that DNA into an empty cell to produce a custom-made, artificial creature. The projects undertaken by Venter's for-profit arm, Synthetic Genomics, Inc. (SGI), particularly in the biofuels area, are considered in Chapter 8. Through the efforts of Venter and four of his competitors, the latter entities discussed in Chapter 8, soon human-made genes will serve as the design components, the building blocks for a new, biology-based technology, having significant commercial uses. Altering the metabolic pathways of microorganisms to engineer living factories will enable businesses to transform feedstocks, such as algae and sugarcane, into a myriad of different molecules used in a wide variety of energy, pharmaceutical, and chemical applications.

Chapter 9 analyzes the policy aspects of writing life.[3] Focusing on the safety of experiments and the risks of terrorists gaining access to the new technologies, the chapter analyzes the concerns about what happens if things go wrong, whether by accident, misuse, or nefarious use. At present, the scary biosafety and biosecurity fears raised by synthetic biology opponents seem farfetched. The chapter briefly discusses the problems posed by the patent system, namely, the possibility of "unfair" monopolies resulting from the granting of intellectual property rights in synthetic organisms, processes, and products.

Those fearful of synthetic biology and the possibility of its unknown, negative, even devastating, consequences, harken back to Mary Shelley's tale about the fateful curiosity and ambition of Victor

Frankenstein,[4] a mythical scientist who could not withstand the lure of discovery. Heedless of the consequences, Frankenstein created new life that returned to destroy him.

In contrast to naysayers, I offer a much more optimistic conclusion regarding synthetic biology and its potential benefits for humankind, finding it, on balance, socially desirable. For me, Pandora opening her box, generating hope, not woe, serves as the better analogy. Relying mainly on self-regulation by the scientific and business communities, my recommended policy framework would guard against governmental overregulation that could create a huge barrier to innovation. The forward momentum of synthetic biology research and commercialization is hard to reverse or contain. However, because of the possible risks, as detailed in Chapter 9, a need exists for ongoing hazard appraisals and the development of enhanced risk assessment tools.

Although synthetic biotechnology holds considerable social and economic potential, offering the likelihood of becoming a major business force this century, at present, it may prove difficult to translate research discoveries into commercially viable applications outside the laboratory. These practical problems ought to temper both the promise and the perils of synthetic biology.

Notes

1.   J. Craig Venter Institute, Press Release, "J. Craig Venter, Ph.D. to Receive National Medal of Science from President Obama," September 18, 2009. For an overview of Venter's career, see Andrew Pollack, "His Corporate Strategy: The Scientific Method," *New York Times*, September 5, 2010, Business Section, 1.

2.   Quoted in Sara Lin, "Craig Venter's Hangout," *Wall Street Journal*, March 5, 2010, W10.

3.   The Presidential Commission for the Study of Bioethical Issues, *New Directions: The Ethics of Synthetic Biology and Emerging Technologies* (Washington, DC, December 2010), 111–74, concluded that neither new federal laws and regulations nor new oversight bodies were currently needed because synthetic biology poses few risks in its early development stages, but recommended that government agencies scrutinize the technology more carefully to minimize risks and foster innovation. For a contrasting view, see, e.g., Erosion, Technology and Concentration (ETC) Group, *Extreme Genetic Engineering: An Introduction to Synthetic Biology* (ETC Group, January 2007), 23–51.

4.   See generally, Mary Wollstonecraft Shelly, *Frankenstein*, curr. ed. (London: Penguin, 2006).

Part I
Background

2

The Beginnings
of Earthly Life

This chapter surveys how earthly life began. There are three broad theories of biogenesis, how original life began.[1] The first theory posits that life originated by chemical self-assembly.

Under the panspermia hypothesis, the second theory, life came to earth in the form of already viable bacteria or microbes. Bacteria are a member of a large group of one-cell microorganisms having cell walls, but lacking organelles (specialized structures in a living cell) or an organized nucleus. Microbes are a type of bacterium causing disease or fermentation. The bacteria or microbes, according to the many versions of the second hypothesis, came from space in the form of extra-terrestrial dust from comets (celestial objects that generally congregated in the outer solar system beyond Neptune), asteroids (small celestial bodies that orbited the inner solar system), or meteorites (small particles of matter in the solar system that reached the earth's surface without being vaporized, and then dissipated). Under the right conditions, these bacteria or microbes could have evolved to become dinosaurs and ultimately, humans.

Under the superbugs theory, the third and most likely approach, life began inside the earth, deep in the oceans, where geothermal activity created cauldron-like conditions. Under the superbugs theory, the first biolife lived on the ocean floor, far removed from the sun's rays. Early life totally depended on chemical energy spewing forth from underwater volcanic vents in the earth's interior.

Before analyzing the origins of earthly life in this chapter, we need to ask: what are the basic components of physiological life? Many contemporary scientific discussions of what constitutes being alive focus on the ability to metabolize, reproduce, and evolve.[2] Metabolic properties connote the use of energy to create order out of disorder and to maintain and generate information, structure, and organization.

Life, however, can exist in the absence of any neurological activity, for instance, some species of microbes, provided the creatures have an enclosure, a cell membrane. With living things having the informational ability to reproduce themselves, life is derived from other living things of its own kind. Finally, life has the ability to evolve through the survival of the fittest.

Earth was a barren place when its surface cooled and solidified some 4.6 billion years ago. Then, life began on earth some 3.8 billion years ago. Evidence from some of the planet's oldest rocks, specifically in the remote regions of West Greenland, supports this theory.[3] Rocks in West Greenland are 3.7–3.9 billion years old. Small fossils, so-called micro-fossils, were long ago disaggregated and destroyed by heat and pressure. What is left behind are the chemical fingerprints of bacteria or microbes.

Chemical Self-Assembly

A primordial soup may have turned chemistry into life.[4] Under chemical self-assembly theory of life, simple chemicals came to life through spontaneous generation. Some type of molecular self-assembly led to the formation of life. The primordial soup required water mixed with sustainable substances, then exposed to an energy source to drive chemical reactions.

The chemical self-assembly found support in the 1953 experiments of Harold Urey, an American chemist and a future Nobel Prize winner for the discovery of deuterium, and Stanley Miller, then a graduate student.[5] Miller filled a glass flask with methane, hydrogen, and ammonia plus water. He sealed the apparatus, but omitted oxygen from the chemical mixture on the theory that oxygen was not around in precursor phase to the origin of life (more technically, the prebiotic phase of life). To simulate sunlight, he passed an electric spark through the mixture. In one week, the water, which slowly turned reddish brown, contained amino acids. As discussed in Chapter 3, amino acids are organic chemicals, formed when carbon and other elements link together. Serving as the building blocks of proteins, they are basic ingredients of earthly life.

The notion of some type of primordial soup, chemical self-assembly, as the beginning of earthly life, seems, however, unlikely.[6] Scientists today do not think the early earthly atmosphere resembled the gases in Miller's flask. Methane and ammonia were unlikely to have been present in abundance. Amino acids are easy to make under a variety of conditions. Although amino acids serve as the building blocks of proteins, the key ingredient of all living things, a big difference exists between building blocks and an assembled structure. The watery soup probably could not link together amino acids to form molecules called peptides. A protein consists of a long peptide chain. Moreover, it is nearly impossible for the amino acid sequences to link together in the requisite order. Proteins rest on specific amino acid sequences. A random, uncontrolled energy injection will not accomplish the needed result. In addition to numerous specialized proteins, life requires a number of other molecules, specifically, DNA and RNA, discussed in Chapter 3, that must be brought together simultaneously.

Panspermia Hypothesis

Turning to the second theory, the panspermia hypothesis,[7] meaning seeds everywhere, asserts that living organisms exist throughout the universe and develop wherever conditions are favorable. Among the many variations of this hypothesis, the concept rests on extraterrestrial dust contained in comets, asteroids, or meteorites landing on earth. These types of debris originated as leftovers from the formation of the solar system.

Scientists have sliced space-dust particles, uncontaminated by the earth's atmosphere, into small slivers, less than one-tenth the thickness of human hair. Data has demonstrated that amino acids are present in some meteorites, bits of asteroids that landed on earth. For instance, a meteorite, containing amino acids, landed in Murchison, southeastern Australia in 1969.[8]

Researchers have also detected glycine, the simplest of the amino acids, in comet bits brought back in 2006 by the U.S. National Aeronautics and Space Administration's space probe Stardust.[9] In January 2004, the Stardust spacecraft flew through the tail of dust and gas of the comet Wild 2. When the probe returned to earth two years later, it parachuted collected samples to the ground for scientists to analyze. Comets are believed to preserve material from the early solar system, largely unchanged during the past 4.5 billion years. Scientists found glycine embedded in the aluminum foil of the spacecraft's collecting apparatus. They confirmed that the glycine came from the comet, not from contamination.

Comets and asteroids that landed as meteorites on earth may possibly have delivered vast quantities of life. The tiny particles were rich in the seeds of life, containing organic compounds, among other elements. Carbon is the key element found in organic compounds. Possessing a unique chemical property, carbon atoms can link together to form a variety of extended molecule chains, such as proteins. Large comets made of ice and rocks likely contained organic compounds. Furthermore, the amino acids the comets, asteroids, and meteorites carried may have survived a devastating landing on earth. Researchers have shown that these amino acids not only withstood the impact but also were transformed into a new compound.[10] The simple amino acids fused together to form peptides, more complex molecules with the peptides linked together to form proteins, still larger building blocks for earthly life. Possibly, larger molecules grew from life's simplest building blocks. However, as with the chemical self-assembly theory, life requires a number of other molecules, beyond amino acids, peptides, and proteins.

Superbugs Theory

Third, and most likely explanation, is the superbugs theory.[11] This hypothesis looks downward in the earth for life's origins. Rather than searching the cosmos, life may have originated deep underground in rocks, or more likely, in hot, deep ocean locations, with the subsequent direction upward, not downward.

As a result of volcanic activity, chemicals from deep inside the earth spewed into the seas. A system of deep-sea hot water, technically, hydrothermal, volcanic openings (or vents), hot springs, formed on the ocean depths.

Thus, earthly life may have originated in a deep, underwater surface, without sunlight. The first microbes drew chemical energy from the water and minerals trapped in the rocks. Clustered around the thermal ocean vents, these organisms may have lived on noxious hydrogen sulfide gas, an inorganic chemical compound, that erupted from the volcanic vents or somehow they ingested carbon dioxide, nitrogen, and hydrogen. They drew enough energy to divide and reproduce, perhaps once every thousand years.

If, as seems likely, that life ascended from the depths of the seabed or perhaps in the porous rock beneath the seabed, the first living organisms did not require: sunlight for their primary energy source; oxygen; or organic material, such as carbon. They obtained what they needed from the hot springs, rocks, and carbon dioxide dissolved in the water.

These early microbes withstood the repeated heavy bombardment, the cosmic impact of rubble of rocks and ice, after the solar system was formed, that sterilized the earth's surface. Without an ozone shield, ultraviolet light would have also proved deadly. Below the surface, however, conditions were more likely stable and more hospitable to early life. A vast area existed to catalyze chemical reactions. The microbes tapped huge reserves of chemical and thermal energy.

These heat-loving, technically thermophilic, microbes that withstand temperatures of upwards of 238 degrees Fahrenheit, well above boiling point of water, have stiffening agents in their membranes to keep them from melting. They build their cell proteins from different kinds of amino acids than human cells do, thereby constructing strongly bonded protein chains that do not collapse in extreme heat.

Over the eons, the first earthly life spread horizontally across the seabed or porous rocks deep in the earth. Then, following a geological upheaval, such as an earthquake or a volcanic eruption, they were stranded in a cooler region on the earth's surface. When the bombardment of comets and asteroids ceased 3.5 billion years ago, some microbial life could have survived outside its protective hiding places deep in the ocean or the earth.

On the earth's surface, nearly all the microbes died because their rigid membranes, and thus their metabolism, could not function at the lower temperatures. A lucky mutant possessing a more flexible membrane survived and multiplied.

When microbial life reached the earth's surface, the survivors could take advantage of the sun, another energy source. From an initial, primitive form of photosynthesis, the microbes evolved chlorophyll, a green pigment that allowed them to trap sunlight and use it to drive photosynthesis, a chemical reaction that converted carbon dioxide and water into chemical energy, while they excreted oxygen as a waste product.

As the cosmic-bombardment abated and the earth cooled, cells in the form of colonies of green slime spread across the entire planetary surface. The growing surface microbes formed stromatolites, domed structures, the remnants of which are found today in rocks throughout the globe.[12] The tiny organisms built up the stromatolites, layer-by-layer, at about one-half millimeter a year, over thousands and thousands of years. They used photosynthesis to collect energy from the sun, but likely secreted a sticky coating to shield them from deadly ultraviolet radiation.

Gradually, the oxygen produced by the primeval organisms built up in the earth's atmosphere, supplanting the noxious gases, such as hydrogen sulfide. The microbes raised the atmospheric level of oxygen from less than 1 percent to its current 21 percent so that plants, animals, and humans could breathe an oxygen-rich air mixture. By creating a layer of ozone in the upper atmosphere, oxygen also protects life from the sun's ultraviolet rays, by screening out this lethal radiation. With the protection of the ozone layer, life on the earth's surface developed and diversified into more complex organisms at first, followed by dinosaurs, primates, and finally humans. The slow task of the early life oxygenating the atmosphere made possible multicellular life.

Recent Discoveries Support the Superbugs Theory

Recent discoveries, supporting the superbugs theory, point to a threefold division of earthly life into eukaryotes, prokaryotes, and archaea.[13] Eukaryotes, such as animals, plants, and fungi, consist of one or more cells containing a membrane-covered nucleus to hold its genetic instruction set in a form that can be read and copied. Genes represent packets of information transferred from one generation to the next during the process of reproduction.

Prokaryotes, such as bacteria, are organisms without nuclei in their cells. Their chromosome is a single circular molecule tacked on to an organism's cell wall.

Archaea, meaning ancient in Greek, an early form of earthly life, comprise a third category. Archaea and prokaryotes went their separate ways hundreds of millions of years ago. The archaean genetic material that controls the mechanism used to convert genetic information into proteins are more like those in eukaryotes rather than in prokaryotes. Eukaryotes and prokaryotes have undergone substantial genetic changes over time, while archaea have evolved far more slowly.

Originally, archaea were thought to be rare and confined to extreme environments. They are now known to be abundant and widely distributed over the earth.

As a result of the identification of the complete genetic code of a seafloor microbe we know that archaea represent a distinctive life form. In 1996, a team led by J. Craig Venter, then president and chairman of the Board of Trustees of The Institute for Genomic Research (TIGR), sequenced the genome—an entire gene set—for *Methanococcus jannaschii*, one type of archaea.[14] This work, as Venter notes, at the time, "[R]epresents the scientific equivalent of opening a new porthole on Earth and discovering a wholly new view of the universe."[15]

M. jannaschii's life is remarkably hardy. It flourishes 8,000 feet deep in the Pacific Ocean, at crushing pressures 245 times greater than at sea level. It was discovered in 1982 above a hydrothermal vent chimney, a cylindrical mineral formation on the sea floor through which hot subterranean fluids empty into the ocean, at ultra-high water temperatures of about 185 degrees Fahrenheit. Without the direct or indirect impact of sunlight and without organic carbon as a food source, it lives by converting carbon dioxide as its carbon source and hydrogen, expelled by the volcanic vents, into small amounts of methane, a hydrocarbon, the main component of natural gas, to generate its cellular energy.

Although TIGR had already published the genetic maps of a bacterium and a second microorganism, as discussed in Chapter 4, *M. jannaschii* was the first archaean ever sequenced. Fifty-six percent of its 1,738 genes were unlike anything known in eukaryotes or prokaryotes, a staggering result, confirming that archaea represent a separate branch on the evolutionary tree of life. "[S]ome of the best matches," of the other 44 percent, Venter noted, "were to human genes and yeast genes."[16] However, each gene without a known counterpart represented a new protein with unknown functions. "That organism [*M. jannaschii*]," Venter recalled was responsible for "stimulating...my thinking, on the energy front."[17] But other things beckoned for Venter, namely, reading the genetic code and sequencing the human genome.

The revelations about the beginning of earthly life and the origins of organisms leave us with many unanswered questions beyond the grasp of scientists, more in the realm of theologians and philosophers. We ponder: are living things more than organized matter? The dimensions of the human experience cannot, in my view, be explained by mere physiological or biochemical analysis. Even if we possess a better understanding of life's genesis, the purpose of earthly existence remains unanswered. Leaving aside our search for coherence and meaning, without which life for most of us seems pointless and empty, we next turn to see how molecular biology elucidated the chemical basis of life.

Notes

1.   I have drawn on Paul Davis, *The Fifth Miracle: The Search for the Origin and Meaning of Life* (New York: Simon & Schuster, 1999), 69–96, 163–86, 221–44, and *Origins: How Life Began*, Broadcasting System, Nova

Transcript Public, September 28 and 29, 2004, http://www.pbs.org/wgbh/nova/transcripts/3112_origins.html (accessed August 14, 2009).

2. See, e.g., Lee M. Silver, *Remaking Eden: How Genetic Engineering and Cloning Will Transform the American Family* (New York: Avon, 1997), 21–24; Francis Crick, *Life Itself: Its Origin and Nature* (New York: Simon & Schuster, 1981), 49–62. See also Mildred K. Cho et al., "Ethical Considerations in Synthesizing a Minimal Genome," *Science* 286, no. 5447 (December 10, 1999): 2087–90, at 2089.

3. S. J. Mojzsis et al., "Evidence for Life on Earth before 3,800 Million Years Ago," *Nature* 384, no. 6604 (November 7, 1996): 55–59.

4. See generally, Davies, *Fifth Miracle*, 81–96.

5. Stanley L. Miller, "Production of Amino Acids under Possible Primitive Earth Conditions," *Science* 117, no. 3046 (May 15, 1953): 528–29. See also Stanley L. Miller, "The Prebiotic Synthesis of Organic Compounds as a Step toward the Origin of Life," in *Major Events in the History of Life*, ed. J. William Schoff (Boston, MA: Jones and Bartlett, 1992), 7–17.

6. Davies, *Fifth Miracle*, 87–93.

7. Ibid., 143–61, 221–27. See also F. H. C. Crick and L. E. Orgel, "Directed Panspermia," *Icarus* 19, no. 3 (July 1973): 341–46; Crick, *Life Itself*, 15–16, 141–53; Robert Olby, *Francis Crick: Hunter of Life's Secrets* (Cold Spring Harbor, NY: Cold Spring Harbor Laboratory Press, 2009), 359–62.

8. Davis, *Fifth Miracle*, 226–27. See also David A. J. Seargent, *Genesis Stone? The Murchison Meteorite and the Beginnings of Life* (The Entrance NSW, Australia: Karagi, 1991), 5–17, 88–170.

9. Jamie E. Elsila, Daniel P. Glavin, and Jason P. Dworkin, "Cometery Glycine Detected in Samples Returned by Stardust," *Meteoritics & Planetary Science* 44, no. 9 (September 2009): 1323–30; Don Brownlee et al., "Comet 81P/Wild 2 Under a Microscope," *Science* 314, no. 5806 (December 2006): 1711–16. See also Kenneth Chang, "From a Distant Comet, a Clue to Life," *New York Times*, August 19, 2009, A24; J. Oro and A. Lazcano, "Comets and the Origin and Evolution of Life," in *Comets and the Origin of Life*, ed. Paul J. Thomas, Christopher F. Chyba, and Christopher P. McKay (New York: Springer, 1997), 3–27.

10. B. T. Liu et al., "Simulation of Comet Impact and Survivability of Organic Compounds," *Proceedings of the 15th American Physical Society Topical Conference on Shock Compression of Condensed Matter* (2007): 1391–94; P. Ehrenfreund et al., "Astrophysical and Astrochemical Insights into the Origin of Life," *Reports on Progress in Physics* 65, no. 10 (October 2002): 1427–87, at 1457–61; J. G. Blank and G. H. Miller, "The Fate of Organic Compounds in Cometery Impacts" (21st International Symposium on Shock Waves, July 20–25, 1997), Paper 8180. See also Oleg Abramov and Stephen J. Mojzsis, "Microbial Habitability of the Hadean Earth during the Late Heavy Bombardment," *Nature* 459, no. 7245 (May 21, 2009): 419–22.

11. Davies, *Fifth Miracle*, 163–83. See also James K. Fredrickson and Tullis C. Onstott, "Microbes Deep inside the Earth," *Scientific American* 275, no. 4 (October 1996): 68–73; Karsten Pedersen, "The Deep Subterranean

Biosphere," *Earth-Science Reviews* 34, no. 4 (August 1993): 243–60; Thomas Gold, "The Deep, Hot Biosphere," *Proceedings of National Academy of Sciences* 89, no. 13 (July 1, 1992): 6045–49; John B. Corliss et al., "Submarine Thermal Springs on the Galápagus Rift," *Science* 203, no. 4385 (March 16, 1979): 1073–83.

12. Martin J. Van Kranendonk, "A Review of the Evidence for Putative Paleoarchaean Life in the Pilbara Craton, Western Australia," in *Earth's Oldest Rocks*, ed. Martin J. Van Kranendonk, R. Hugh Smithies, and Vickie C. Bennett (Amsterdam, The Netherlands: Elsevier, 2007), 855–77. See also Martin J. Van Kranendonk, "Volcanic Degassing, Hydrothermal Circulation and the Flourishing of Early Life on Earth: A Review of the Evidence from c. 3490–3240 Ma Rocks of the Pilbara Craton, Western Australia," *Earth-Science Reviews* 74, nos. 3–4 (February 2006): 197–240; H. J. Hofmann et al., "Origin of 3.45 Ga Coniform Stromatolites in Warrawoona Group, Western Australia," *Bulletin of the Geological Society of America* 111, no. 8 (August 1999): 1256–62.

13. The existence of archaea, as a distinctive category of life, was first proposed by Carl R. Woese and George E. Fox, "Phylogenetic Structure of the Prokaryotic Domain: The Primary Kingdoms," *Proceedings of National Academy of Sciences* 74, no. 11 (November 1977): 5988–90; Carl R. Woese and George E. Fox, "The Concept of Cellular Evolution," *Journal of Molecular Evolution* 10, no. 1 (March 1977): 1–6. For a view that early evolution was characterized by the swapping of genes across the three species lines, see Nicholas Wade, "Life Origins Get Murkeir and Messier," *New York Times*, June 13, 2000, D1. There may be a fourth category. See Dongying Wu et al., "Stalking the Fourth Domain in Metagenomic Data: Searching for, Discovering, and Interpreting Novel, Deep Branches in Marker Gene Phylogentic Trees," *PLoS One* 6, no. 3 (March 2011): e18011.

14. Carol J. Bult et al., "Complete Genome Sequence of the Methanogenic Archaeon, *Methanococcus jannaschii*," *Science* 273, no. 5278 (August 23, 1966): 1056–73. See also J. Craig Venter, *A Life Decoded: My Genome: My Life* (New York; Penguin, 2007), 211–13; Kathy Sawyer, "From Deep in the Earth, Revelations of Life," *Washington Post*, April 6, 1997, A1; Curt Suplee, "Imminent Domain," *Washington Post*, September 23, 1996, A3; Jessica Mathews, "The Aliens Among Us," *Washington Post*, September 3, 1996, A15; "Hot Stuff," *Economist* 340, no. 7980 (August 24, 1996): 63–64; Nicholas Wade, "Deep Sea Yields a Clue to Life's Origin," *New York Times*, August 23, 1996, A23; Paul Recer, "Decoding of Microbe's Genes Reveals 'Very Different' Life Form, Researcher Says," *Washington Post*, August 23, 1996, A22; Tim Friend, "1977 Discovery Confirmed as Third Life Form," *USA Today*, August 23, 1996, D2.

15. Quoted in Suplee, "Imminent Domain."

16. Ibid.

17. Quoted in Ross Douthat, "The God of Small Things," *Atlantic Monthly* 299, no. 1 (January/February 2007): 120–25, at 123.

3

The Design for Life

James D. Watson and Francis H.C. Crick's momentous discovery of the double helix structure of DNA in 1953 impacted both science and society. As Watson recounted:

> Contained in the molecule's graceful curves was the key to molecular biology, a new science whose progress ... has been astounding. Not only has it yielded a stunning array of insights into fundamental biological processes, but it is now having an ever more profound impact on medicine, on agriculture, and on the law. DNA is no longer a matter of interest only to white-coated scientists in obscure university laboratories; it affects us all.[1]

This chapter provides an overview of the chemical basis of life—the design for life—as revealed by molecular biology. With a few exceptions, the same DNA-based biochemistry forms the foundation of genetic structure in all life. Because, for our purposes, all living things today use this same genetic structure, they all descend from one original cell that began life. The universal structure points to a common ancestor for all life, a topic considered in Chapter 2.

This chapter focuses on three topics: cells; the chemical composition of deoxyribonucleic acid (DNA) and its structure; and how organisms create proteins. The chapter concludes with a brief discussion on the biotechnology revolution resulting from this new knowledge. Today, biotechnology, using rDNA techniques, involves making modest changes in cells to make them serve commercial and medical purposes. Before we begin this overview, some definitions are helpful.

Genomics is the study of the complete genetic makeup of living organisms, or, more technically, their genomes. A genome consists of all of the DNA contained in the cellular nucleus of a species and represents the species' total genetic information. Therefore, the genome contains all of the hereditary information needed to design, assemble, and operate an organism.

DNA, the genetic text which spells out the recipe for life—the instructions on how to make each species—consists of four chemicals known as base pairs. Similar to the information carried by any sentence, the information carried by the DNA molecule depends on the precise sequence of these base pairs. The paired bases form the rungs of a DNA ladder with two DNA chains linked together so that each base in the pair complements the other.

Genes are long stretches of DNA, the genetic material parents pass to their offspring. Genes, the sequences of DNA that specify how to build proteins, represent the smallest functional unit of the genetic code. As the individual units of information that give each cell its specific biochemical instructions, genes are made of small molecules of DNA like beads on a string. In a gene, the DNA beads are analogous to letters in words.

A gene instructs the cellular machinery how to manufacture proteins. Proteins carry out various tasks and processes within an organism, functioning as the key drivers and mediators of cellular function and biological system activity. Amino acids serve as the building blocks for these all-important proteins, and a large part of the DNA databank stores instructions on how to make proteins from amino acids.

But DNA does not directly instruct the cellular machinery. Instead, another molecule, ribonucleic acid (RNA), carries a portion the DNA's code from the DNA to a ribosome, a piece of this cellular machinery. A ribosome uses the instructions in the RNA to assemble amino acids in the proper order to construct proteins.

DNA is packaged with proteins, among other cellular components, into neat little packages called chromosomes, the microorganic particles in each cell's nucleus. Thus, a cell's chromosomes, consisting of DNA bound to proteins, function as a repository of genetic information and contain the entirety of a cell's genome.

We come full circle to each organism's genome; the total amount of chromosomal DNA spooled in the nucleus of every one of its typical cells. Genomic sequencing determines the exact order of the four chemical subunits, or base pairs, in a DNA strand so that this information can be related to the biochemical activity influenced by the piece of DNA.

The Cell: The Basic Unit of Biolife

The basic living unit of biolife is the cell. It is microscopic in size. Most human cells are less than one-tenth of a millimeter across. Every living thing contains at least one cell, and most living things grow larger by increasing the number of their cells rather than by increasing their cell sizes.

Most cells have a common appearance and inner workings. Each cell is surrounded by a plasma membrane, an ultra-thin skin. Within its skin, the cell has hundreds of thousands of working parts, each localized to a specific compartment and each communicating with the other cellular components.

All cells work in essentially the same way with the same complex molecules and the same type of genetic material, which is read from the same genetic structure. The type of cell, however, depends on which specific genes are activated within it. In the human body, for example, there are about two hundred cell types, all of which are coded from the same DNA. More generally, there are two categories of cells: eukaryotes and prokaryotes. Eukaryotic cells, such as those of humans, generally consist of two separate compartments: the nucleus and cytoplasm. However, in prokaryotic cells, like those of bacteria, the cells are not compartmentalized. Lacking a nucleus, the genetic material of bacteria floats freely in their cytoplasm.

The nucleus, which has its own membrane, sits in the middle of each eukaryotic cell. The nucleus contains all of the cell's genetic material within structures called chromosomes. Simple-cell organisms contain one chromosome, whereas normal human cells carry twenty-three chromosome pairs for a total of forty-six chromosomes, specifically consisting of twenty-two nonsex chromosome pairs and one sex chromosome pair.

The cytoplasm consists of all the cellular material outside the nucleus and inside the plasma membrane. In prokaryotes, there is no nucleus, so the cytoplasm also contains the cell's genetic material. In eukaryotes, the cytoplasm interprets the genetic information flowing from the nucleus and responds to this information by building the various structures that make up the cell. The cytoplasm also relays signals from within the cell, from other cells in the body, and from the external environment to the nucleus. In response to these signals, the nucleus then elicits specific changes in its method of gene expression, the formation of proteins. Each cell's genetic material, its DNA, contains the information required to produce the cell's numerous protein components in the correct numbers and to place each in the right location. In this way, gene expression controls the physical attributes of cells, tissues, organs, and entire organisms.

DNA's Chemical Composition and Its Structure

A DNA molecule contains billions of atoms, linked together in the distinctive form of two coils, twin helical strands connected by cross-links. The now-famous twisted double helix discovered by James D. Watson and Francis H. C. Crick in 1953 is bundled up in a convoluted shape, with two interlocking side chains wound around each other.[2] The arrangement of the atoms along the helical strands of DNA serves as the instruction manual for each living species. Humans share the magical DNA molecule with other earthly life forms, each molded according to their respective DNA instructions. The chemical makeup of DNA and the double helix structure is universal, although each species' DNA is unique.

It is uncertain how the first DNA molecule was formed. The DNA molecule has possessed the ability to replicate itself since the first primitive organisms. Gradually, increasingly lengthy instructions for more complex organisms came into being. Eventually, DNA molecules of sufficient complexity to produce human beings were constructed, containing tremendous amounts of genetic information in comparison with our primitive ancestors.

Genetic information is encoded within each DNA molecule as a series of well-defined, basic units. A nucleotide, the modular component of DNA, represents one basic unit, and each nucleotide contains one of four different chemical bases. These bases are used over and over again and are strung together in various configurations along a DNA molecule. The four nucleotide bases are referred to by the first letter of their chemical names: A (adenine), C (cytosine), G (guanine), and T (thymine). A nucleotide consists of three components: a phosphate group, a sugar, and one of two types of bases, a pyrimidine (C and T) or a purine (A and G). These bases are flat, ring-shaped molecules containing carbon, nitrogen, oxygen, and hydrogen atoms.

The unique characteristics of each living species derive from the specific messages recorded in their respective DNA molecules. The information carried by the DNA depends on the sequence of the chemicals, A, C, G, and T. The language of DNA consists of a linear series of As, Cs, Gs, and Ts written as three-letter words, such as "TGA." Because each three-letter word corresponds to specific genetic information, the message written into the DNA's base pairs is crucial.

The DNA molecules in living cells are enormous compared to other kinds of molecules. A DNA molecule is an incredibly long, but narrow, thread. If the DNA molecules present in each human cell were lined up end to end, they would measure some six feet eight inches (or two meters) in length but only 2.4 nanometers in width. But because their diameter is so narrow, an entire set of DNA molecules fits within the nucleus of each of the one hundred trillion cells in the human body. In short, the genetic information contained in each DNA molecule is written across its length, not its width.

In discovering the basic machinery of life, Watson and Crick deduced from the way DNA's base pairs were linked together and how one strand of DNA could be copied into a complementary strand. This discovery led to a revelation of how cells replicated the DNA in their chromosomes when they get divided.

Watson and Crick found that in forming a double helix with two complementary strands, A always pairs with T and C always pairs with G. The twisted ladder, double helix structure allows a DNA molecule to split lengthwise into two single helix pieces, each of which forms a new complementary piece, replacing its lost partner and passing on the code of life. As Crick noted, "It is this specific pairing between bases on opposite strands that is the heart of the replication process. Whatever sequence is written on one of the chains, the other chain must have the complementary sequence, given by the pairing rules."[3] In short, Watson and Crick "found the basic copying mechanism by which life comes from life."[4]

The double helix design makes the copying of DNA from one generation to the next simple. As the two strands of the double helix pull apart, the nucleotide bases of the single strands gather up their complements from the supply of free nucleotides in the cell's mix of chemicals—the Ts grab the As, the Cs grab the Gs, until each separate strand reconstructs an exact replica of its lost partner, forming two new identical double helixes, each with one new strand and one old strand. It may be helpful to visualize DNA as a zipper-like molecule, with chemical bases denoted by the letters A, C, G, and T as the teeth of the zipper. Because A binds only to T and G binds only to C, unzipping and rezipping creates two identical copies of the original zipper. In this way, an organism's DNA can replicate itself and contain the same information because of the chemical affinities between the letters.

The Formation of Proteins

When a gene is expressed, it is translated into a protein. The protein that the gene expresses together with all the other proteins enables an organism to develop and live. Thus, proteins determine the way cells and entire organisms form and function.

DNA instructs the cellular machinery to make proteins via RNA.[5] RNA, a molecule more ancient than DNA, is used by a cell for multiple purposes in the protein synthesis process.[6] RNA consists of four bases: A, G, C, and U (uracil) (U is similar to T and is used instead of T in RNA), but unlike DNA, RNA is only a single helix.

RNA carries the code from DNA to a ribosome, a piece of cellular machinery in the cell's cytoplasm. In turn, a ribosome uses the instructions on the RNA to create a protein from one of twenty different amino acids. The DNA template directs RNA synthesis in a process called transcription.[7] The RNA template then directs the protein synthesis in a process called translation.[8] Thus, RNA links DNA to proteins. Longer strings of DNA make more complex molecules by linking together more amino acids. In this process, the linear, one-dimensional DNA digital code is translated into a string of amino acids that fold into a three-dimensional functioning protein molecule.

The protein formation process begins with the transcription of the DNA contained in the cell's nucleus. In the transcription process part of the DNA double helix is unzipped and one single-stranded RNA copy, called messenger RNA (mRNA), is synthesized. The mRNA, much like a copied section of DNA, is complimentary to the original section of DNA.

The mRNA reads the protein recipes from DNA in the nucleus and conveys them to mini-molecular factories, more technically ribosomes, located in the cell's cytoplasm where proteins are made. The ribosomes serve as factories for protein synthesis, bringing another type of RNA, transfer RNA (tRNA) into positions where tRNA can read the information incorporated in the mRNA. Each tRNA molecule comes to the ribosome with one type of amino acid stuck to its end. The mRNA carries the instructions so that the targeted tRNA molecule, carrying the designated amino acid, will recognize the exposed bit of mRNA and bind to it. Other tRNA molecules, carrying the incorrect amino acids, will not fit properly into the binding site. As the protein is manufactured, the mRNA strand chugs through a slot in a ribosome, which carries out the instructions bit-by-bit, hooking the tRNAs' cargos of individual amino acids together one-by-one in a specified sequence, until an entire new protein molecule is constructed. The amino acids are lined up and linked together to create a chain that folds to form a protein. When the protein synthesis is complete, the ribosome receives a stop signal from the mRNA and the chain cuts loose from the ribosome.

Proteins are made from twenty different types of amino acids. All twenty possess a common structural element that permits them to be knitted together into strings of proteins. However, the chemical differences allow the amino acids to perform different functions.

Amino acids consist of two parts, a shared backbone and a residue. The shared backbones link the amino acids together, while the unique residues define the specific properties of each amino acid. Some possess an electric charge, positive or negative; some are slim, others, bulky; some are flexible, others stiff. Chemical differences allow amino acids to perform widely differing functions. Some amino acids are best at linking to others; others excel at catalyzing specific types of reactions. Each different sequence of amino acids yields a different protein, and the order and type of the amino acids determine the chemical properties of the proteins they make. However, proteins are more than a linear string of amino acids. The key to a protein's function is how the string folds to produce a three-dimensional configuration.

The cell builds proteins as long chains, thereby connecting one amino acid to another through the backbone of each. A peptide is a short chain of amino acids. A polypeptide is a long chain of usually more than one hundred amino acids. The information required to produce the protein's three-dimensional structure rests in the one-dimensional sequence of amino acids present in its polypeptide chain. The packet of information present along a DNA molecule, a gene, determines the one-dimensional sequence of amino acids.

Proteins become twisted and folded into highly specific three-dimensional structures with specific physical and chemical properties. The majority of amino acids in most proteins are there to fit together with other amino acids thereby ensuring the proper twisting and folding occurs. Thus, each protein achieves its unique shape and form so that it can function properly, with the bumps and cavities in the right places and the correct atoms facing outwards. Once they assume this complex configuration, the proteins become biologically active.

Living creatures are composed mainly of protein molecules that display considerable flexibility in form and function. Proteins, the working parts of a cell, are generally multipurpose tools, with each role performed by a different section of a protein. In other words, proteins multitask. Some play structural roles, becoming cellular structures themselves or serving as precursors to other structures, such as membrane components that surround cells; other proteins help in building the connective tissue between cells; many proteins play key roles in health, regulating metabolism or serving as enzymes, catalyzing a cell's chemical reactions. Some proteins act as readers of genetic information and build new DNA molecules based on the information contained in the old DNA molecules. This flexibility enables proteins to function as the cell's machinery, carrying out almost all the work that goes on inside a cell.

Practical Implications of Molecular Biology Discoveries

The molecular biology insights into the genetic code provided the foundations for a biotechnology revolution. The discoveries of Watson and Crick eventually led researchers to develop the means to sequence and directly alter DNA with a high degree of accuracy. In turn, these rDNA technologies gave rise to the modern biotechnology era.

Using rDNA technologies, first researchers and then businesses learned to cut and paste discrete genetic materials from existing organisms into new DNA sequences.[9] Starting with an organism's unmodified genome, rDNA technologies allow scientists to modify the organism's DNA or combine it with that of another in various ways. For example, genes may be spliced from one organism to another or genes may be forced to mutate for a specific purpose. Inserting, mutating, or deleting a single gene or a few significant genes modifies a cell's processes. Thus, by using rDNA technologies, useful genes may be moved into bacteria or other organisms to create unique gene expressions that were previously unavailable.

In this way, rDNA biotechnology revolutionized medical science by creating new drugs, therapies, and medical diagnostics.[10] For example, scientists utilize genetic engineering to coax microorganisms such as the common bacterium *E. coli* to produce insulin and growth hormone as well as protein pharmaceuticals. Produced by a genetically altered organism, recombinant insulin, obtained by splicing the human insulin gene into a bacterium, represents a marked improvement over pig insulin. Today, recombinant proteins, such as synthetic insulin, produced by genetic engineering of microorganisms are industrially fermented to produce vast amounts of low cost, pure, and potent biotherapeutics.

In agriculture, companies use rDNA technology to develop new varieties of genetically engineered, pest-resistant corn, soybeans, and cotton, among other crops, with seeds genetically engineered to ward off insects.[11] Moreover, these seeds are also genetically engineered to make a crop herbicide tolerant. These types of genetic modifications help farmers save work and enable them to cultivate more land.

Along with the rDNA technology came another technological advance, the perfection of methods for reading the sequencing of DNA. Using new technologies, scientists can accurately determine the exact sequence of A, C, G, and T nucleotides in any DNA sequence, allowing them to map the entire genome of an organism. Ultimately, using these technologies researchers sequenced the entire human genome, led by J. Craig Venter.

Notes

1. James D. Watson with Andrew Berry, *DNA: The Secret of Life* (New York: Knopf, 2003), xiii.
2. J. D. Watson and F. H. C. Crick, "Molecular Structure of Nucleic Acids: A Structure for Deoxyribose Nucleic Acid," *Nature* 171, no. 4356 (April 25, 1953): 737–38; J. D. Watson and F. H. C. Crick, "Genetical Implications of the Structure of Deoxribonucleic Acid," *Nature* 171, no. 4361 (May 30, 1953): 964–67. James Watson recounted the discovery of DNA's structure in *The Double Helix: A Personal Account of the Discovery of the Structure of DNA* (New York: Atheneum, 1968). See Ibid., xi–xii, 45–55. Francis Crick presented his views of the discovery in *What Mad Pursuit: A Personal View of Scientific Discover* (New York: Basic Books, 1988), 62–79; "The Double Helix: A Personal View," *Molecular Biology* 248, no. 5451 (April 26, 1974): 766–69; "How to Live With a Golden Helix," *The Sciences* 19, no. 7 (September 1979): 69. See also Matt Ridley, *Francis Crick: Discoverer of the Genetic Code* (New York: Harper Collins, 2006), 45–57, 59–76; Robert Olby, *Francis Crick: Hunter of Life's Secrets* (Cold Spring Harbor, NY: Cold Spring Harbor Laboratory Press, 2009): 126–95; Robert Olby, *The Path to the Double Helix: The Discovery of DNA* (New York: Dover, 1994): 385–423; Horace Freeland Judson, *The Eighth Day of Creation: Makers of the Revolution in Biology*, exp. ed. (Plainview, NY: Cold Spring Harbor Laboratory Press, 1996), 3–49, 51–123, 125–69. For a detailed analysis of the structure of DNA, see James D. Watson et al., *Molecular Biology of the Gene*, 6th ed. (San Francisco, CA: Pearson/Benjamin Cummings, 2008), 102–27.
3. Crick, *What Mad Pursuit*, 62–63.
4. Francis Crick, letter to Michael Crick, March 17, 1953, quoted in Ridley, *Francis Crick*, 77.
5. Crick told his quest for the protein synthesis mechanism in *What Mad Pursuit*, 89–101, 109–12, 113–36. In *Genes, Girls and Gamow: After the Double Helix* (New York: Knopf, 2002), Watson offered his views on the search to unravel RNA. For a technical discussion, see F. H. C. Crick, "On Protein Synthesis," *Symposia for the Society for Experimental Biology* 12

(1958): 138–63; *DNA Makes RNA Makes Protein*, ed. Tim Hunt, Steve Prentis, and John Tooze (Amsterdam, the Netherlands: Elsevier Biomedical Press, 1983). Watson summarized the role of RNA in *DNA*, 69–78.

6. For a detailed analysis of the structure of RNA, see Watson, *Molecular Biology of the Gene*, 127–32.

7. For a technical discussion of the transcription mechanism, see Ibid., 377–412.

8. For a technical discussion of the translation mechanism, see Ibid., 457–517.

9. Watson provided an overview of rDNA research in *DNA*, 87–96. For a technical background on the methods of creating rDNA molecules, see James D. Watson et al., *Recombinant DNA*, 2nd ed. (New York: Scientific American, 1992), 63–77.

10. Watson summarized the impact of biotechnology on medicine in *DNA*, 113–31. For technical background on rDNA in medicine, see Watson, *Recombinant DNA*, 453–70.

11. Watson presented an overview of genetically modified agriculture in *DNA*, 131–33, 135–51. For technical background on the genetic engineering of plants, see Watson, *Recombinant DNA*, 273–92 and the generation of agriculturally important plants, Ibid., 471–78.

Part II
Reading Life

4

Reading the Genetic Code: Sequencing the Human Genome

In March 2000, after a series of failed negotiations between the leaders of the public and private programs to sequence the human genome, President William Jefferson (Bill) Clinton intervened and forced the two sides to collaborate.[1] As a result of the political pressure, renewed conversations occurred between the two rival programs.

The public program, the International Human Genome Sequencing Consortium, was led by Francis S. Collins, Ph.D., the then-director of the National Human Genome Research Institute (NHGRI) at the NIH. After a group of scientists convinced the federal government to fund an extensive effort to sequence (or map) the entire human genome, the Human Genome Project officially kicked off on October 1, 1990.[2] The program would ultimately cost about $3 billion in federal government funding. As the public consortium evolved over the years, it came to consist of a group of academic centers financed largely by the NIH and The Wellcome Trust of Great Britain, then the world's largest non-profit medical research foundation. Five major centers included Baylor University (College of Medicine), Washington University, Whitehead Institute at MIT, Joint Genome Institute at the U.S. Department of Energy (DoE), and the Sanger Center in the United Kingdom, the latter funded by The Wellcome Trust of Great Britain.

J. Craig Venter, president and chief science officer of Celera, a for-profit firm, led the private effort. Although long involved mapping the genomes of various organisms, in September 1999, almost nine years behind the public effort, Venter turned his attention to the human genome.

These two groups competed to be the first to map the entire human genome, but on June 26, 2000, the human genome sequencing race ended in an anticlimactic tie. President Clinton presided over a White House ceremony, the first time in history that a major scientific advance was announced there, in which Collins and Venter jointly announced the mapping of the human genome.[3] Both sides projected a spirit of cooperation after an acrimonious rivalry.

Sequencing a genome means determining the order of all the chemical building blocks of an organism's DNA. The printed result of a computer-generated genome is a string of As, Cs, Gs, and Ts, the four chemical components of the genetic code. Computers annotate the text of a genome, where its genes start, pinpointing the on and off gene switches found in DNA regions between the genes. By comparing the newly discovered DNA sequences with the genes of known function whose sequences exist in databanks, software can identify the likely role of many newly discovered genes. In short, powerful computers and sophisticated software capable of decoding the order of chemical letters in long strings of DNA make possible the study of genomics, an organism's complete genetic makeup.

As of June 2000, however, the human genome was not fully mapped. Instead, the results were presented as rough drafts that included about 90 percent of the genome, but featured many gaps. Both sides had only a rough draft of the order and number of the four chemicals, A, C, G, and T, present in human chromosomes; more work was needed. Over the next six months or so, the two teams raced to fully sequence the human genome, and in February 2001, the two groups published their findings separately.[4]

In mapping the human genetic code, the two projects had learned the correct identity and order of the three billion DNA subunits that make up the genes in the human body. The sequences, as published, were actually half genomes (more technically, a haploid genome) and represented only one set, or twenty-three chromosomes, of the forty-six pairs of chromosomes in every cell. When the final number of human genes was tabulated, researchers concluded that human chromosomes contain between twenty-three thousand and thirty thousand genes per set.

Subsequently, in September 2007, researchers at the J. Craig Venter Institute (Venter Institute), together with collaborators at The Hospital for Sick Children in Toronto and the University of California, San Diego (UCSD), published the genomic sequence of one individual, J. Craig Venter, the leader of the for-profit sequencing team. Venter's genome covered both sets of his chromosome pairs (more technically, a diploid genome), or all forty-six of his chromosomes, with one set representing the total genetic information inherited from his father and the other set from his mother.[5]

This chapter focuses on the efforts of Venter to read DNA, first at the NIH, then at TIGR (largely funded by a for-profit firm, HGS), and finally at another for-profit venture, Celera. This chapter also provides an overview of Venter's two organizational structures after he left the NIH, his team's scientific successes, and the reasons his two post-NIH arrangements ended despite his research achievements culminating in sequencing the human genome. The chapter then briefly concludes with a discussion of the practical applications of genome sequencing.

Venter at the National Institute of Health

After a stint as a faculty member at the State University of New York in Buffalo, in 1984 J. Craig Venter, a Ph.D. in biochemistry, landed at the NIH in suburban Washington, DC.[6] At first, Venter, recruited by the NIH's Institute for Neurological Disorders and Strokes, worked on locating and decoding a protein in human brain cells, but he found the progress exasperatingly slow.

In 1986, when he learned about a machine that could "read" DNA by shining lasers on its dyed letters, he flew to meet its co-inventor, Michael W. Hunkapiller, Ph.D., in California. Although the NIH would not pay for a prototype DNA "reader" (more technically, a DNA sequencer) manufactured by Applied Biosystems, Inc. (the company Hunkapiller had joined), Venter found the funds for the automated DNA sequencing machine from another project, his biological warfare detection work for the U.S. Department of Defense.[7]

A DNA sequencing machine, like the one Venter purchased, determines the order of chemical subunits (technically, nucleotide bases) in genetic material. The new technology pioneered by Applied Biosystems simplified and accelerated the process of sequencing letters in any stretch of DNA. In order to function, however, DNA sequencing machines require the generation of large amounts of the relevant DNA, which rested on the development of the polymerase chain reaction (PCR) process[8] invented in 1983. Using the PCR process, researchers copy a small sample of DNA and amplify its signal, thereby enabling its sequence to be read far more easily. Today, robots churn out vast quantities of the heat-resistant polymerase enzyme used in the PCR process, so that DNA can be replicated accurately and rapidly for high-speed sequencing machines.

In order to determine the exact sequence of a molecule of DNA, Venter's automated DNA sequencer passed an electric current through a gel medium containing a mixture of DNA. In the mixture were DNA pieces of varying length, synthesized from an original DNA sequence. But while each piece began with the same letter, or nucleotide, a piece's termination point was determined at random. For example, if the original sequence was "TAGCAG," then the mixture would contain sequences such as "T," "TA," "TAG," "TAGC," and so on, all the way up to the full length of the original, here "TAGCAG." In this way, the mixture represented every possible complimentary sequence which could be synthesized using "T" as a replication starting point. The mixture of DNA flowed through the gel when an electronic potential (technically a bias) was applied thanks to DNA's natural negative charge. The rate of flow of a particular piece was dependent on the DNA's charge to mass ratio, with more massive DNA fragments tending to move slower.

Venter used fluorescent dyes to tag these moving DNA fragments, with the dye color depending on the last letter of sequence. Therefore, "T," "TA," and "TAG" would all be different colors, even though they share some common letters. As each DNA fragment ran through the gel toward the positive electrode, it passed in front of a laser beam that activated the fluorescent dye. The color emitted by each fragment indicated the identity of the terminating DNA nucleotide and was recorded using an electric eye (technically, a photomultiplier tube), obviating the need for excruciating manual data-entry. Because shorter sequences moved through the gel faster, they would reach the eye sooner than longer sequences. In this way, each DNA length would reach the eye in successive order ("T" before "TA" before "TAG"), providing a direct readout of the last nucleotide in that sequence ("T" then "A" then "G"). The computer automatically interpreted the various colors, identified the different base pairs, and recorded the nucleotide sequence and its length. Using this information, Venter could determine the exact order of the nucleotides in any piece of DNA, or any gene.[9]

Not all DNA encodes genes, however. Much of an organism's chromosomal DNA contains "junk" DNA, or DNA that does not code for any protein. Rather than hunting for genes and attempting to sequence them, Venter came up with a key discovery while at the NIH. Instead of using DNA directly, he collected mRNA that mirrored the DNA's coding. When a cell is about to use a gene, enzymes in the cell convert the DNA sequence into mRNA that helps direct the making of proteins. Significantly, the mRNA only would mirror genes, not junk. Because each mRNA molecule picks up the gene's code and only the gene's code directly from DNA, using mRNA enables researchers to deduce the sequence of DNA in the gene, effectively isolating it from the junk DNA. However, mRNA was (and is) unstable and fragile. To solve this problem, Venter inserted the RNA into bacterial cells programmed to clone junk-free complementary DNA (cDNA), using mRNA as a template.

Venter then took thousands of small bits of cDNA produced in the bacteria and used automated sequencing machines to read the cDNA's genetic instructions.[10] By taking a short segment (some 300–500 letters) from a piece of cDNA, known as an expressed sequence tag, Venter could quickly and easily spell out the fragment's sequence using the new automated machine. These express sequence tags, sufficient to identify the gene's message and its uniqueness from the other genes expressed in the cell, could verify the presence of a gene and could subsequently be used to pin down the gene's location on a chromosome, and when isolated the tags could be used to roughly identify individual genes.

The advantage of Venter's approach was speed; however, the information generated from express sequence tags was not comprehensive. At this point, Venter's purpose was not to sequence the entire human genome, but rather to identify individual genes present in humans. While the purpose of each gene could not be ascertained, his approach did facilitate the suggestion of a gene's function with clues derived from the similarity to known genes in other organisms. It also aided the identification of previously undiscovered genes of unknown function. By using expressed sequence tags, researchers could compare these sequences in with those public databases and see if they were similar to known genes from other organisms.

Despite its limited scope, Venter's rough and ready method generated more information than many expected. The sequenced short regions of cDNAs were often sufficiently distinctive that the function of the full gene could be guessed. Depending on the gene, even with only one-tenth to one-third of the total sequence, Venter could often identify the entire gene. Using computer databases to search all known genes, Venter compared the newly discovered genes with known human or other genes, and frequently, he was able to use sufficiently similar genes to guess at new genes' functions in humans.

At the NIH, Venter's work, over the years, gradually shifted to molecular biology, specifically, gene research. Making a commitment to DNA sequencing as his major scientific pursuit, he sought funding from the NIH's National Center for Human Genome Research (NCHGR) in late 1989, and then again in 1990 and 1991, but was repeatedly rejected. Venter also became embroiled in a battle at the NIH over the patenting of human gene fragments, before the function of these partial gene sequences could be known.[11] Following a series of disputes over funding and patents with James D. Watson, then the director of the NCHGR (originally, the Office of Human Genome Research) at the NIH, and other NIH leaders, Venter departed the NIH in July 1992.[12] In his capacity as director of the NCHGR, Watson had decided not to fund Venter's partial gene sequencing approach, favoring instead the large-scale, complete genome mapping that would have culminated in the genomic sequencing of unedited DNA. As directed by Watson, in the early years of the federal genome project through 1994, NIH policy excluded cDNA sequencing and Venter's expressed sequence tags. This policy decision opened the door to private funding for cDNA fragments. As one academic public policy expert put it, "With the door open to capturing the commercial value of sequence data one gene at a time, private sector cDNA sequencing efforts took off."[13]

After leaving the NIH on good terms, Venter embarked on two private ventures that greatly expanded his DNA sequencing efforts. These brought him scientific glory and riches, but anguish resulted from the corporate world and the controls it placed on him, especially with respect to the publication of his team's research findings.

Venter and Human Genome Sciences, Inc.

In 1992, Venter chose to continue his DNA research outside the public sector. He and his team conducted genomic research at his own independent institute, TIGR. At TIGR, Venter pursued increasingly ambitious projects on DNA sequencing. These research projects were largely funded by a privately held, for-profit, startup company, HGS, which sought to turn TIGR's scientific findings into marketable drugs. This section analyzes the TIGR–HGS organizational structure, Venter team's scientific discoveries, and the eventual divorce of TIGR and HGS.

Organizational Structure

After leaving the NIH, Venter sought to establish his own institute where he could conduct research for academic purposes without facing worried investors. The realization of his goal, TIGR, was subsequently incorporated in July 1992. Its mission as stated in its articles of incorporation: "[I]s to perform scientific research relating to biological and computer sciences, including, but not limited to, mapping the human, animal, and plant genomes by means of gene sequencing and other techniques...."[14]

Meanwhile, the then-chairman of the board of directors of Health-Care Investment Corp., a leading venture capital company in the healthcare field, Wallace H. Steinberg, wanted to get his hands on the DNA sequencing technology developed by Venter.[15] Steinberg hoped that Venter's shortcut gene-sequencing technique would accelerate the discovery and production of drugs, but after leaving the NIH, Venter had no desire to run a business focused on drug discovery and production.

Nevertheless, together with the incorporation of TIGR, Steinberg established HGS, a new biotechnology company formed by Health-Care Investment Corp., as the for-profit commercial partner of TIGR to market the institute's discoveries, with HGS initially financing TIGR's research. Using the gene-hunting techniques developed by Venter, HGS would commercialize TIGR's findings. Although not an HGS officer, Venter received a 10 percent interest in HGS,[16] which went public in December 1993. HGS subsequently floated additional shares publicly in September 1995 and March 1997.

Initially, HGS agreed to provide TIGR with a ten-year, $70 million grant; however, the deal changed in October 1992 when TIGR entered into a Research Services Agreement and an Intellectual Property Agreement with HGS.[17] Pursuant to these two agreements, along with a Lease Funding Agreement in March 1993 and a subsequent agreement entered into April 1993, HGS committed to provide $85 million to TIGR over a ten-year period, ending September 2002. In return, these agreements required TIGR to disclose to HGS all significant developments relating to information or inventions discovered at the institute, and further provided that HGS would own, on a royalty-free basis, all of TIGR's interest in inventions and patent rights arising out of the institute's efforts during the term of the agreement, subject to certain exceptions.

The ability to quickly publish discoveries was key to Venter. Therefore, the original agreements incorporated a three-tier approach to publication: HGS was allowed six months after discovery to select genes to turn into drugs, and if selected, there would be a six-month delay before TIGR could freely publish its data regarding these selected genes. However, if any of the selected genes showed promise as drug blockbuster that could generate more than $1 billion per year in revenue, HGS would have an additional eighteen months to develop them.[18] But after the lapse of HGS' six-month or two-year periods, Venter's team could freely publish their discoveries.

In May 1993, to fund its arrangement with TIGR, HGS gave Smith-Kline Beecham PLC (SmithKline) exclusive rights to TIGR's and HGS' discoveries for useful genes in the human genome. SmithKline committed to pay $125 million for the exclusive rights to commercialize therapeutic, vaccine, and diagnostic products based on human genes discovered by TIGR and HGS and received an equity stake in HGS. In turn, HGS would receive royalties from products developed and sold by SmithKline based on the discoveries, would be able to co-promote certain products, and retained certain rights, for example, to the use of genetic data for applications in gene therapy.[19] Pursuant to the agreement, SmithKline paid an initial $55 million in 1993, with the remaining $70 million to be parceled out as license fees, option rights, and milestone payments.

In June 1996, HGS and SmithKline modified their collaboration and licensing agreement. This modification allowed HGS and SmithKline to independently designate certain potential therapeutic proteins for exclusive development and commercialization. The modification also allowed the HGS/SmithKline venture to enter into collaborative agreements with other drug companies, thereby facilitating the speed and breadth of product research and development. Thereafter, in June and July 1996, the HGS/SmithKline venture entered into a series of collaborative arrangements with Schering-Plough Corp., Synthelabo SA, and Merck KGaA. These agreements provided HGS/SmithKline with at least $140 million to be shared equally and received over an initial five-year research period in the form of license fees and product development milestone payments in exchange for the rights to research, develop, and commercialize therapeutic products based on HGS/SmithKline human gene technology. Royalties and research-support payments from the new collaborations went to HGS and SmithKline received certain product marketing rights.[20]

Despite the infusions of cash, quarrels soon developed between TIGR and HGS regarding the publication of data.[21] Venter bickered with HGS and its lawyers over when he could publicly release TIGR's discoveries, which the firm felt would undermine its commercial position. Venter complained, "My own creativity and that of other scientists here [at TIGR] was being stifled."[22]

In an attempt to resolve the publication dispute, TIGR, HGS, and SmithKline, in July 1994, agreed to contribute a number of partial human cDNA sequences to the Human cDNA Database. This database was accessible to academic scientists and researchers at nonprofit institutions and government agencies who signed access agreements,[23] but was not available to entities or persons engaged in commercial activities.

Pursuant to the Human cDNA Database Agreement, HGS took the lead in disclosing its research data. In July 1994, HGS opened its databank to academic scientists under terms that protected the firm's proprietary interests. Then, in September 1994, Venter announced that scientists would receive access to the genetic data accumulated by TIGR and a large amount of data HGS garnered from its own in-house research following the publication of a report on his research over the two prior years in a scientific journal.[24] Achieving this goal took an additional year, and in September 1995, Venter opened a large part of TIGR's vast storehouse of genes to university and government scientists.[25] In so doing, researchers obtained access to about three hundred thousand partial gene sequences, without any restrictions. Venter did this through two media; first, scientific journal supplement that featured a paper by Venter and his team describing the molecular sequences to what were believed to be a least one half of all human genes,[26] and second, on the internet to academic researchers. Additionally, academic and government researchers could study another forty-five thousand partial sequences that were proprietary to HGS only if they agreed to certain conditions, including giving HGS the first right to commercialize drugs developed from the information. Drug and biotechnology companies, among other commercial interests, however, could only tap into the gene databanks by working through HGS.

The enormity of the data provided was overwhelming. It had taken Venter two years to generate, clean up, and analyze the data, and consultants to the journal needed an additional year to review the massive amount of information contained solely in the paper.

Scientific Discoveries

Venter and his team did not sequence the human genome overnight. Their sequencing efforts went through three phases: first, a bacterium; second, the fruit fly; and third, human. The first was done during Venter's HGS phase; the latter two, while Venter was at Celera Genomics.

Sequencing the Genome of a Bacterium

With private funds, in May 1995, Venter won the race to sequence the genome of a humble bacterium, the first genome of a free-living organism. Previously, in August 1994, the NIH had turned down his grant proposal, saying that the task seemed impossible because the pioneering Venter–Smith decoding method was unworkable.

To speed up the genome sequencing, in 1993, Venter had turned to a new recruit, Hamilton O. Smith, Ph.D., a Nobel laureate then at Johns Hopkins University School of Medicine. In 1978, Smith was the co-recipient of the Nobel Prize in physiology or medicine for his work on bacteria restriction enzymes used to cut DNA in specific places. Restriction enzymes serve as tiny chemical scissors that cut DNA at a precise point. Researchers use these enzymes to chop up pieces of DNA so they can be studied or recombined with other organisms' DNA. Aware of Smith's prowess, Venter invited him to be an adviser to TIGR.

In December 1993, Smith proposed to Venter and his TIGR team a strategy for sequencing the genome of a bacterium by shotgunning an organism's entire genome. The controversial shotgun method, a bottom-up technique, differed radically from the generally used, but laborious mapping method, a top-down approach.[27]

The top-down mapping method involved first making a map of the chemical signposts along the DNA chain, using known genes to place these signposts. The technique then sequenced all of the DNA in between the signposts. The sequenced signpost pieces served as landmarks, keeping researchers oriented. Scientists sequenced each DNA fragment between the landmarks using automated sequencing machines that recorded each sequence's order of letters. This process was repeated as more landmarks were identified until an entire genome was sequenced.

In contrast to the mapping method, the whole genome shotgun sequencing method is simpler. First, the genome is broken into random chunks of DNA that are then sequenced using an automated sequencing machine. These random sequences are then fed into a computer that arranges the sequences in the correct order using complicated software algorithms.

The Venter–Smith shotgun technique began by blasting a genome into tiny, random pieces. Using a kitchen blender, Smith shattered an organism's DNA in millions of small fragments, then inserted the fragments into bacterial hosts (technically, vectors) used to replicate and amplify each fragment. This collection of doctored molecules prepared for automated sequencing constituted a DNA library.

Venter's team then used software programs to fit together the millions of pieces of the enormous DNA jigsaw puzzle by comparing them and looking for overlaps between the fragments. It was no simple matter to reassemble the fragments in the correct order, like interlocking pieces of a jigsaw puzzle, as they exist in the organism's genome. The sheer length of a genome coupled with the four-letter DNA design meant that same or similar lengths of DNA were often present in more than one chunk of DNA. Furthermore, the pieces were sequenced without prior mapping of their chromosomal positions. Because the pieces were randomly generated, however, researchers could identify overlaps with the help of high-speed computers and special algorithms programmed to search for matching regions.[28] If the left end of one piece of DNA had the same sequence of letters as the right end of another piece of DNA, they were overlaid to form a single longer sequence. By repeating the process, piece by piece the computers stitched together millions of pieces of DNA into an entire genome.

While he had developed a mechanism to sequence the human genome, Venter was not ready to begin this ambitious project. To take a key step in understanding how biochemistry creates life, Venter turned his focus from the human genome to that of bacterium, much to HGS' discontent. His first test of the genome shotgun method involved a tiny microbe, *Haemophilus influenza*, that can cause serious illnesses, including ear infections and childhood meningitis. The humble bacterium possesses all the tools needed for independent existence.

Announced at a scientific meeting in May 1995 and published in July that year,[29] the *Haemophilus influenza* genome showed for the first time the genetic apparatus needed for self-contained existence. The sequencing and computer reconstruction represent the first decoding of a complete genome, the complete set of genes needed for life, of a free-living organism.

Although the shotgun methodology did not identify the function of the genes, the Venter–Smith discovery changed the world of microbial genomics. The sequenced genome offered crucial insights into the biology of bacteria. Scientists could see everything a living cell needed to grow, survive, and reproduce itself. They poured over the data, identifying each gene, trying to determine each gene's function, how genes worked with one another, and how human genes were related to similar genes in other creatures. In annotating the genome, researchers compared each new gene against databases with previously discovered genes in other living things and were able to assign functions to more than half of the genes, although 40 percent could not be accounted for.

While nature reuses most microbe genes in complex mammals, including humans, the *Haemophilus influenza* genome is small; it includes only 1.8 million letters of genetic code and slightly more than 1,700 genes. In 1995, it was unclear whether Venter and his team could ramp up shotgun sequencing for a species having a far larger genome, such as humans. However, with a full catalog of an organism's genes in hand, Venter could start with a gene and search for its function by comparing it with genes of known function in databases.

To prove the discovery was not a fluke, Venter's team also sequenced the genome of a second living single-cell organism, the *Mycoplasma genitalium*,[30] that is responsible for reproductive-tract infections. With 580,087 base pairs, the *Mycoplasma* has the smallest genome—its genetic instruction set—of any known free-living organism. Venter chose this tiny organism because it would help determine the set of genes needed for life. And by sequencing this genome, Venter's team discovered that an organism may need as few as about three hundred genes to function.

As discussed in Chapter 2, in 1996, the Venter team subsequently sequenced the genome of *Methanococcus jannaschii*, an archaean type of living thing, with some 1,700 genes, found in deep sea hot vents. Genomic research provided a way of looking at the origins of life and its initial divisions among prokaryotes, eukaryotes, and archaea.

Despite successfully sequencing the *Haemophilus influenza* genome, Venter was dissatisfied with his business arrangements with HGS. According to Venter, HGS tried to hold up publication of the scientific paper reporting this discover, asserting the dispute "made it clear to me we would have to spend money on lawyers instead of science."[31] The rift between TIGR's and HGS' goals gradually widened, eventually leading to a permanent split.

Corporate Divorce

June 1997 marked the corporate divorce of TIGR from HGS. Neither TIGR nor HGS came to need each other. The relationship had become more problematic, especially after Steinberg's death in July 1995, and TIGR's and HGS' goals became increasingly incompatible. TIGR began seeking grants from other parties, including the NIH and the United States Department of Energy, and as TIGR successfully obtained grants from these sources, it became less dependent on HGS funding. TIGR's research focus also changed from the human genome to those of bacteria and other organisms.

HGS also went in a new direction. Originally, HGS planned to rely primarily on the human DNA sequences produced by TIGR. However, as Venter and his team turned to more basic science, focusing on the genomes of bacteria and other simple creatures, and HGS faced growing competition from rivals, the firm changed its course and downplayed TIGR's role. After duplicating TIGR's sequencing and computer systems, HGS began churning out hundreds of thousands of pieces of human DNA code daily and patenting the most important new genes that were useful to HGS' goal of developing breakthrough drugs.

Ultimately, in June 1997, TIGR severed formal ties with HGS,[32] and HGS relinquished rights to any future work done by TIGR. The settlement permitted TIGR to publish its new genetic data immediately. In return, TIGR gave up the right to $38 million in funding from HGS that would have been paid over the next five-and-one-half years, and also agreed not to enter into any commercial agreements for the next four years on specific therapeutic proteins and associated diagnostic tests being developed by HGS.

Free from HGS, TIGR could operate more like an academic research institution, funded with some $30 million in HGS stock which Venter had put into the institute.[33] The separation of the institute from its commercial partner enabled TIGR to gain tax-exempt status, helping facilitate the openness of genome research to scientists, as TIGR put a significant amount of new genetic information into GenBank, the public genetic database administered by the National Center for Biotechnology Information at the NIH.

By mid-1997, HGS had more than 60 of its own DNA sequencing machines and a database of more than 1.25 million different human

DNA sequences. After HGS had mostly completed its scientific mission of discovering all human genes, it turned to drug discovery. Wanting to turn scientific findings into marketable drugs, HGS transformed itself from a science-based company to a product-development firm. In 1997, HGS had more than one dozen drugs in pre-clinical trials, with the start of human clinical trials for two drugs expected in 1998.

Venter and Celera Genomics Corp.

Six months after being free from HGS, Venter cultivated a new funding source. In February 1998, Michael W. Hunkapiller, co-inventor of the automated DNA sequencing machine originally used by Venter at the NIH and now president of the PE Biosystems division of Perkin-Elmer Corp., showed Venter the prototype components of a new Applied Biosystems Prism 3700 DNA Analyzer, a sequencer five times faster and more highly automated than any other sequencer on the market. In place of the unwieldy and slow gel, the new sequencer machines used very thin, multicapillary tubes (technically, high-throughput capillary system) in which DNA fragments were rapidly sorted by size and type thereby eliminating manual gel pouring, allowing much faster sample processing by automating the loading of samples, and reducing the amount of human intervention involved. When the fragments flowed out the end of the capillary tube, laser signaling triggered each fluorescent dye and the resulting order of base pairs of genetic letters was determined. Moreover, the new machine could work 24/7 with little monitoring. With this technology, Venter and Hunkapiller soon concluded that it would be possible to sequence the human DNA within three years, at a cost of some $300 million.[34]

Venter's wife protested his linking up again with the corporate world. As one author related their conversation[35]:

> To his wife, he was sounding unbalanced again, like the old Craig who couldn't tell the difference between a sound idea and a stupid one. "Mike's machine is unbelievable, Claire!" he exclaimed. "They want to start a new company, with me as its head. With enough of the new machines, I think we could do it. We could get the entire human genome!"
> Claire Fraser kept quiet, as she had long ago decided to do in these situations: just let him rave for a while and he'd come to his senses on his won. But he kept pressing her, following her around the house. She sat down in the sunken living room and turned on the television. He grabbed the remote and flicked it off. "So what do you think?" he said. "I really want to know."

"Are you insane?" she answered. "It's been less than a year that you got free of Haseltine [the CEO of HGS]. Do you want to give up your science, and have some company trying to control you again?"

"It's not like that. For god's sake, it's *Mike*. We've worked together for years. We understand each other. We have the same vision."

"Yeah?" she said. "Well, what about this Tony White [the CEO of Perkin-Elmer]? Who's he?"

"Tony White is not important," Venter replied. "Trust me. He's not going to be a problem."

In the end, Venter went ahead with Perkin-Elmer. His wife's views proved prophetic.

Organizational Structure

In May 1998, Venter and TIGR joined Perkin-Elmer Corp. to decipher the human genome, hopefully within three years.[36] Perkin-Elmer was a longtime maker of scientific instruments and a leading supplier of systems for life science research. Its Applied Biosystems division was the preeminent manufacturer of machines used to sequence DNA, dominating the market.

Using the Applied Biosystems' breakthrough technology and the sequencing strategies pioneered by Venter's team, the new firm, Celera, came to operate a vast genomics sequencing facility. The new company sought to become the world's definitive source of genomic and related biological and medical information. Commenting on the new venture, Venter stated, "By linking technologies that have been used by TIGR scientists with Perkin-Elmer's genetic analysis technologies, we pave the way for a new era of post-genomic discovery."[37]

Celera also sought to build the scientific expertise and the informatics tools needed to extract biological knowledge from the genomic data. When Venter announced in May 1998 that Celera would beat the Human Genome Project by four years, the bombshell turned a leisurely race into a fierce pursuit. Venter spurred the international consortium to move faster; the Human Genome Project feared being upstaged by a privately owned rival financed by the company that made the DNA sequencing machines used by the consortium. Thus, with the emergence of Celera, the public effort exhibited a new sense of urgency to map the human genome.

Perkin-Elmer financed Venter's new venture. In return for an infusion of $330 million in the second quarter of 1998, Perkin-Elmer obtained ownership of about 80 percent of the new, independent firm, Celera, with Venter, TIGR, and others owning the balance. Venter's ownership equaled 5 percent; TIGR received another 5 percent.[38] Venter would serve as the new company's president and chief scientific officer. He resigned as president of TIGR and was replaced by his then-wife, Claire Fraser, Ph.D., but continued as chairman of the Board of Trustees of TIGR.

As part of the deal, Venter made it clear he wanted to sequence the human genome and offer both free, public access to the data every three months and publish a major paper on his team's discovery. Perkin-Elmer agreed to his twin demands[39] without restricting how other researchers could use the data, including not seeking intellectual property protection on others' discoveries using Celera's data. Celera would only seek database protection to bar other firms from selling the information contained in its database.[40] In a subsequent SEC filing, the firm stated[41]:

> The Celera Genomics Group[CGG] has adopted a policy to make available to the research community for free, the basic reference human sequence information generated by its sequencing and assembly efforts. [CGG] intends to make available this information on a quarterly basis in the form of unordered consensus human sequence data in excess of specified lengths.
>
> The data that [CGG] releases publicly will be available, in a searchable form, via its web site. The ultimate form of data release will be affected by, among other things, the evolution of intellectual property and [CGG's] assessment of the likelihood that other organizations may seek to obtain [CGG's] data and resell it to their own customers...
>
> After completion of sequencing, [CGG] expects to release a detailed ordered consensus human genome assembly.
>
> [CGG] believes that disclosing consensus sequence data will not affect the value of its information products and services and will encourage researchers to use its data and ultimately become [CGG's] customers.

The new company began settling into its facility in August 1998. Celera's scientists were busy in 1999, with its data center becoming operational at the beginning of that year, and the firm pursued its revenue-generating strategy.

Celera initially expected to derive its revenues primarily from its data customers. The firm offered its information products, including a variety of databases, on a multiyear subscription basis, without the payment of royalties on certain (but not all) products made using Celera data. Its online information and discovery system, Celera Discovery System, provided users access to the firm's genomic and related biological information. In quick succession, Celera signed five-year database subscription deals with Amgen Inc., Pharmacia & Upjohn, Inc., Novartis Pharma AG, and Pfizer,[42] among other firms. In return for unrestricted early access to its data, each of these companies agreed to pay Celera a minimum of $5 million per year for five years.[43]

After Perkin-Elmer sold its analytical testing instruments unit for $425 million to EG&G, Inc. (which renamed itself PerkinElmer Inc.), in May 1999,[44] the original Perkin-Elmer became PE Corp., basically a life sciences and genomics firm. The restructured firm focused its operations on two units. First, the PE Biosystems Group (subsequently renamed the Applied Biosystems Group in November 2000), which developed and marketed instruments-based systems, including the leading brand of DNA sequencing machines, reagents (chemicals used to create chemical reactions to detect, measure, or produce other substances), and software to the life science industry and research community. And second, the Celera Genomics Group, Venter's unit.

As part of the recapitalization of the new PE Corp., in 1999, the company established two new classes of common stock, so-called tracking stocks, that separately tracked the performance of its PE Biosystems Group and its Celera Genomics Group businesses. PE Corp. took PE Biosystems Group and Celera Genomics Group, neither of which had its own board of directors, public thorough these special classes of tracking shares. In May 1999, for each share a stockholder owned in the old Perkin-Elmer firm, he, she, or it would get one share of the PE Biosystems Group and one half share of the Celera Genomics Group. Then, in November 2000, PE Corp. became Applera Corp., a combination of Applied Biosystems, which also adopted its new name in November 2000, and Celera.

During the genomics investment bubble, the price of Celera shares quickly skyrocketed. Its stock climbed from its opening price of $13.50 a share on April 28, 1999, reaching $276 and closing at $252 per share (the pre-stock-split equivalent of more than $500 a share) on February 25, 2000, a mere ten months later. At that time, Celera's stock market value reached a staggering $14 billion.

As guided by its parent, PE Corp., Celera began to expand from genomics to the development of therapeutic drugs. The initial step focused on moving into proteomics, an understanding of proteins, their form, expression, and patterns of interaction. PE Biosystems had nearly completed a new mass spectrometer that promised to make possible the analysis of proteins associated with disease onset or progression and, therefore, could be potential targets for therapeutic intervention or markers for disease detection and progression on an industry-size scale. With PE Biosystems providing technological support, Celera had a good chance to become one of the world's leading sources of protein data.[45] To finance Celera's continuing research efforts in mapping the human genome and its move into functional genomics, with an emphasis on proteomics, PE Corp. floated a secondary offering of Celera Genomics Group's common stock in March 2000, at the height of the genomics investment bubble. The offering proved a spectacular success with Celera netting a cash infusion of nearly one billion ($943.3 million) dollars.

One month after reaching its per share peak of $276, Celera stock fell to $92.12 a share by March 31, 2000; dropping to $36.12 by December 29, 2000 and $27.40 by December 28, 2001. The shares slid after investors realized that Celera's data business, facing competition from the free information in public databases, would never grow fast enough to warrant the sky-high valuation the shares enjoyed in early 2000. Also, on March 14, 2000, President Clinton and U.K. Prime Minister Tony Blair issued a joint statement to the effect that the raw fundamental data on the human genome "should be made freely available to scientists everywhere."[46] Celera's stock nosedived $56 dollars that day, along with that of other genomics companies, as fears spread that some new U.S. Government policy might undermine potentially valuable patents on genes. The next day, a White House spokesman told reporters that the administration supported the patenting of genes.

Scientific Discoveries

With the Perkin-Elmer deal in hand, Venter set his sights on the human genome in competition with the public program. Wanting to demonstrate the operational capabilities of Celera's facilities, specifically, the efficiency of the whole genome shotgun method for sequencing, assembling, and ordering DNA, Venter needed reassurance before spending more than $100 million to sequence the human genome.[47] Therefore, beginning in May 1998, Venter turned his attention to an old laboratory favorite because it breeds quickly and easily: the fruit fly, *Drosophila melanogaster*. The fruit fly has a longer genome than the two bacteria Venter's team successfully mapped, at 1.2 billion letters long representing some 13,600 genes.

Although using the state-of-the-art 24/7 sequencing technology supplied by PE Biosystems and its own sophisticated, internally developed information technology, during the first half of 1999 Celera faced technical problems resulting from sequencing machines failure rates. Ultimately, newly designed robot sequencers successfully ran 24/7, and by September 1999, Celera's research team unveiled the fruit fly's genome. Their findings were published in March 2000,[48] and Venter released the fruit fly data to GenBank with no restrictions on its use, despite the dismay of PE's management.[49]

In annotating the fruit fly genome that was assembled from three million pieces of DNA, researchers found a significant correspondence between human beings and the fruit fly, at least on the genetic level. Some 175 of 289 genes known to cause diseases in humans had analogs in the fruit fly, leading scientists to postulate that genes in the fruit fly could be manipulated as one step to devising remedies for human diseases.

The sequencing of fruit fly, with some 13,600 genes, provided a test of both the shotgun's ability to decode a large genome and its accuracy. With Celera using the new computerized equipment to sequence the fruit fly's genetic code, Venter became ever more confident of the power of shotgun sequencing. Convinced that the public project's approach was needlessly slow and expensive, Venter turned to the human genome. When Venter began sequencing of the human genome in September 1999, the public project had decoded about 25 percent of the three billion letters of the genome.[50]

With the bugs in the sequencer machines worked out, Venter led a factory-style operation that featured three hundred new style, fully automated sequencers, each the size of a small refrigerator and costing $300,000. Celera typically ran three hundred sequencers simultaneously, each day decoding fifty to one hundred million base pairs of high-quality DNA.[51] Using vast processing power and a huge memory, Venter created the world's biggest computer then in civilian use to process the raw data derived from the sequencers. Celera also designed and built new software tools for its advanced computer, able to handle tens of millions of DNA fragments daily, thereby coupling sequencer automation with a new type of computational analysis.

Besides its own efforts, each day the public program deposited new human DNA sequences into GenBank, the publicly accessible database. To save time and money, Celera helped itself to the public data to supplement its own. By using the information generated by the public project to sort the DNA pieces into smaller piles corresponding to known locations, Venter's team could, with the problem somewhat simplified, put together the genome map more quickly. But while the public data helped pin down Celera's random reads produced by the shotgun method to a specific region of the genome, some of the data was of low quality and degraded Celera's assembly and ordering.

Over the years, Venter had also deposited considerable human data in the GenBank much like the public project, but Celera did not make any of its post-September 1999 data available to the public consortium. When the Human Genome Project announced it would compete with Celera in mapping the human genome, Celera's parent blocked Venter from releasing any human data piecemeal, until his team published a scientific paper on the entire human genome. Venter did not fight the order, agreeing not to publicly make available Celera's human genome data prior to its appearance in a journal.[52]

In late 1999 and early 2000, Venter's and the public teams raced to sequence, assemble, and order every bit of human genetic information. In the end, the two groups shared the credit by presenting separate human genome drafts simultaneously before a White House gathering in June 2000.

The Denouement

Venter's plan for Celera to sell access, on a subscription basis, to the human genome map and other information to pharmaceutical companies, biotechnology firms, research institutions, and universities ultimately failed. Although Celera eventually had more than fifty commercial and academic subscribers around the world, in January 2002, Venter was nevertheless ousted by Applera's board as Celera's president and chief scientific officer.[53] In its search for profits, Applera turned Celera away from genomics, with the hope that a drug development business plan would offer the potential of extraordinary profits.

Celera's business plan initially focused on selling access to detailed gene maps of humans and other organisms, to various institutional subscribers, gathered and organized in a user-friendly database. Venter wanted Celera to be the "Bloomberg of biology,"[54] the definitive source for genomic information, with value-added databases it developed, including the complete sequence of the human genome, a database of subtle genetic differences in humans (technically, single nucleotide polymorphisms), links to biological and disease information, and tools to analyze the data, as well as the genomes of two important research organisms, the fruit fly and the mouse (the latter genome Celera completed in April 2001).[55] As a one-stop shopping source for genetic information, these analytical tools would enable subscribers to mine the data for its biomedical value.

However, Celera's original business plan faltered. The plan did not work as expected, because most of the information became publicly available primarily through the federal government's efforts to provide free access to genome sequences. After entering into an exclusive marketing and distribution agreement with Applied Biosystems to distribute its human genomic, biological, and medical information in April 2002, by April 2005 Celera was unable to make a profit by selling this information, and was forced to discontinue its genome subscription business. Ultimately, Celera decided to put the information into the public domain by transferring it to the GenBank.[56]

Long before 2005, however, Celera had turned to a new business strategy. Beginning in 2001, its parent, Applera, began to shift Celera to developing diagnostics and therapeutics. As it shifted focus, Celera made two significant moves. In July 2001, Celera teamed up with its parent, Applera, Applied Biosystems, and another sibling firm, Celera Diagnostics, a diagnostic tests joint venture formed in April of that year between Applied Biosystems and Celera,[57] to begin its own drug discovery efforts. Also, that month, Applera announced the Applera Genomics Initiative, a $100 million project with, among other objectives, the goal of commercializing products derived from information obtained through an analysis of variations in the human genome.[58] The project sought to identify and validate genetic variations and mistakes in the genome, single nucleotide polymorphisms in genes that make individuals unique in various traits such as hair, skin and eye color, height, and weight, and genes that are linked to disease susceptibility and the risk of disease, drug side effects and impact on the efficacy of treatment. This phase in its genomics strategy, Applera hoped, would transform Celera from a company helping others discover drugs to one that made its own discoveries.

To capitalize on Celera's human genome discovery foundations, Applera sought to transform that information into novel therapeutic and diagnostic products. Seeking to implement its new business plan, the Applera subsidiaries worked together to develop new products, including diagnostic tests and drugs, based on the new information, particularly about key genetic differences. It was expected that discovering and focusing on genetic variations, specifically medically relevant differences among individuals, would help identify disease-related gene associations and monitor how genes are expressed and proteins formed. Celera would then develop and market products based on these discoveries.

In a move to accelerate Celera's expansion into the drug discovery arena, in November 2001, the firm acquired Axys Pharmaceuticals Inc., a small molecule (chemical-based) drug company, to expand its drug discovery and development efforts.[59] Appelera hoped the acquisition would enable Celera to build a therapeutics discovery and development business by combining its existing genomics capabilities and its developing proteomics potential with Axys' medicinal and structural chemistry and biological competencies. Although Venter was instrumental in this acquisition, he wanted to keep his focus on unraveling genomes and providing genetic databases, not leading pharmaceutical research and development. When Celera switched gears, Venter's days at Celera were numbered, and two months after the acquisition of Axys, Appelera terminated Venter.

The Medical and Commercial Significance of Reading the Genetic Code

The promise of mapping the human genome is far reaching. It is hoped that knowledge of the complete sequence of human DNA will enable development of better diagnostic techniques and drugs to combat human health problems. Today, armed with the information provided by the human genome, researchers continue to tease out the secrets of human health and disease.

Mapping the human genome has helped researchers learn how the human body works and what may go wrong. Sequencing has played a major role in the discovery of genetic causes of some conditions, including some common ailments and the treatment of these genetically based diseases.[60] Moreover, by sequencing human DNA and studying families for aberrations in the genetic code associated with diseases, researchers hope to create more accurate diagnostic tests and develop more potent drugs with fewer side effects. With more detailed knowledge of how genes express themselves in healthy humans, researchers and physicians hope to understand the mechanism of many more afflictions.

Genetic sequencing also may impact individualized preventive medicine in innovative ways. Often times, individual genetic variations are a tip-off of a predisposition to certain diseases. A few genes may significantly increase the likelihood of getting certain diseases, while other genes may pose modest risks individually but together may negatively impact on health. Through genetic testing and the development of general genetic profiles it is currently possible to foresee a person's future health problems, thereby shifting medicine from a treatment mode to a predict-and-prevent mode and expanding the scope of preventive medicine.

Personalized genetic diagnostic tests are also expected to contribute to a more personalized approach to treatment.[61] It is hoped that physicians will use sophisticated, whole-genome tests designed to provide access to a complex, comprehensive body of information to help select the most effective drugs, if needed, with fewer negative side effects. This approach will likely benefit patients with more customized care, reduced illness length, and better and earlier treatment results.

The new genetics may also lead to new and better therapies. Physicians will be able to use the newly acquired genetic knowledge to develop drugs to directly counteract the underlying molecular problems in human ailments, such as cancer. Furthermore, a new understanding of bacteria will allow vaccine makers to design genetically crippled bacteria to serve as superior live vaccines. These new vaccines may replace antibiotics that face bacterial resistance to them.

Research into diagnostics and therapeutics has thus far been slow, however, and corporate failures litter the way. The medical promise of the human genome may take much longer to be fulfilled. One example, DeCODE genetics, Inc. (DeCODE), a pioneering firm that used the relatively homogenous gene pool in the Icelandic population as the testing ground for detecting disease-causing mutations, filed for bankruptcy in November 2009.[62] Before the bankruptcy, the company compiled a large database containing the genetic profiles and health records of many Icelanders and scanned the database to look for links between certain genes and diseases. It led the race to identify the genetic causes of diseases, such as schizophrenia, diabetes, and prostate cancer.

But the finish line was much further away than expected. In addition to financial mismanagement, DeCODE floundered because the genetic nature of most disease is extremely complex. Many diseases do not have simple genetic pattern of causation. Humans are not that easily decoded. Our bodies are full of variations and quirks. Many genes work together in mysterious ways, together with the environment in which genes operate. In other words, a person's fate is usually not tied only to one's genes. Environmental and psychological factors also impact an individual's physical and mental health as much as (if not more than) a person's genes alone. In fact, genetic mutations associated with a disease alone can account for only part of the overall incidence of the disease. DeCODE found that mutations alone were specifically responsible for too few cases to support the commercial development of diagnostic tests or blockbuster pharmaceuticals, resulting in the firm's failure.

Because scientists need to understand how a specific gene causes a disease before they can develop effective drugs, discovering treatments became a lengthy process. Research takes time, but along with failures, there are also some successes. In March 2011, the U.S. Food and Drug Administration approved HGS' drug for lupus.[63] The drug blocks a protein that stimulates certain cells that are part of the immune system. HGS, a pioneer in genomics and the analysis of human DNA, discovered the gene for that protein. If successful, the drug would become the first new treatment in more than forty years for the autoimmune disease in which the defense system against pathogens attacks the body's own tissues and one of the first drugs to arise directly from genomics research.

Looking decades into the future, we may witness the development of individualized medicine, or tailoring drugs to patients depending on specific variations in their DNA sequences. A doctor might read a person's genome card, a computer chip the size of a postage stamp, showing his or her complete biological software code and holding a person's entire genetic code, including the genes that make him or her vulnerable to specific diseases. If he or she becomes ill, a physician could analyze his or her genes, conclude that a certain drug is best, and track how the body responds to treatment.

A significant degree of realism is needed, however, when discussing personalized medicine. Pharmacogenomics, custom-made medicines, based on one's genotype is likely decades away, and even knowing a person's genetic makeup may not be enough. In humans, multiple genes lead to genetic variations from cell to cell, tissue to tissue, and organ to organ. In other words, one's DNA is a dynamic, evolving text.

Genomic sequencing efforts, human and otherwise, also led to the development of comparative genomics. By comparing the DNA sequences of various organisms, scientists have gained a deep insight into the differences between the major branches of life and how these differences may have evolved. Currently, researchers are working to trace the tree of evolution back to a common ancestor of all life.

After his January 2002 termination as a Celera officer, Venter planned to spend more time fulfilling his role as chair of the board of TIGR. He quickly regrouped. He noted, "All my life I have been a dreamer and a builder, and this was not the time to stop. I decided that it would be easier to start from scratch, just as I had all those years ago when I worked at the NIH. I decided to move forward, to at least strive to do something new that could have an even bigger impact than sequencing the human genome."[64] True to his vision, Venter went on to found a nonprofit research empire and a new for-profit firm.

Notes

1. J. Craig Venter, *A Life Decoded: My Genome, My Life* (New York: Penguin, 2007), 298–319; James Schreeve, *The Genome War: How Craig Venter Tried to Capture the Code of Life and Save the World* (New York: Knopf, 2004), 341–65.

2. James D. Watson with Andrew Berry recounted the beginnings of the Human Genome Project in *DNA: The Secret of Life* (New York: Knopf, 2003), 166–73, 178–79. For background on the political and historical events behind the Human Genome Project, initiated by the DoE, but was outmuscled by the NIH financially and intellectually, see Robert Cook-Deegan, *The Gene Wars: Science, Politics, and the Human Genome* (New York: Norton, 1994), 79–185. John Sulston and Georgina Ferry, *The Common Thread: A Story of Science, Politics, Ethics and the Human Genome* (Washington, DC: Joseph Henry, 2002) 62–80, 105–9, 149–259, provide a glowing tribute to the Human Genome Project and a bitter critique of J. Craig Venter's efforts. For a summary of how the human genome was sequenced, see James D. Watson et al., *Recombinant DNA: Genes and Genomes—A Short Course*, 3rd ed. (New York: W.H. Freeman, 2007), 249–307.

3. Celera Genomics Corp. (Celera), Press Release, "J. Craig Venter, Ph.D. and Chief Scientific Officer, Celera Genomics Remarks at the Human Genome Announcement," June 26, 2000. See also Nicholas Wade, "Genetic Code of Human Life is Cracked by Scientists," *New York Times*, June 27, 2000, A1; Rick Weiss and Justin Gillis, "Teams Finish Mapping Human DNA," *Washington Post*, June 27, 2000, A1; Scott Hensley, "Celera, U.S. Group Discuss Publishing Genome Findings," *Wall Street Journal*, June 15, 2000, A12; Frederic Golden, Michael D. Lemonick, and Dick Thompson, "The Race is Over," *Time* 156, no. 1 (July 3, 2000): 18–23.

4.   J. Craig Venter et al., "The Sequence of the Human Genome," *Science* 291, no. 5507 (February 16, 2001): 1304–51; International Human Genome Sequencing Consortium, "Initial Sequencing and Analysis of the Human Genome," *Nature* 409, no. 6822 (February 15, 2001): 860–921; Celera, Press Release, "Celera Genomics Submits Human Genome Manuscript for Publication," December 6, 2000; Celera, Press Release, "Celera Genomics Publishes First Analysis of Human Genome," February 12, 2001. See also Venter, *Life Decoded*, 320–24; Nicholas Wade, "Genome Analysis Shows Humans Survive on Low Number of Genes," *New York Times*, February 11, 2001, 1; Nicholas Wade, "Long-Held Beliefs are Challenged by New Human Genome Analysis," *New York Times*, February 12, 2001, A20; Nicholas Wade, "Genome's Riddle," *New York Times*, February 13, 2001, F1; Rick Weiss, "Life's Blueprint in Less Than an Inch," *Washington Post*, February 11, 2001, A1; Chris Adams, "Rival Genome Researchers Give Glimpse of Findings, Commend Parallel Efforts," *Wall Street Journal*, February 13, 2001, B6; Scott Hensley, "Celera's Genome Anchors it Atop Biotech," *Wall Street Journal*, February 12, 2001, A3.

5.   Samuel Levy et al., "The Diploid Genome Sequence of an Individual Human," *PLoS Biology* 5, no. 10 (October 2007), http://www.plosbiology.org (accessed May 5, 2010) and J. Craig Venter Institute, Press Release, "First Individual Diploid Human Genome Published by Researchers at J. Craig Venter Institute," September 3, 2007. See also Rick Weiss, "Mom's Genes or Dad's? Map Can Tell," *Washington Post*, September 4, 2007, A1; Nicholas Wade, "In the Genome Race, the Sequel is Personal," *New York Times*, September 4, 2007, F1; Brian Vastag, "The Venter Decryption," *Science News* 172, no. 10 (September 8, 2007): 147–48; Nicholas Wade, "Scientist Reveals Secret of Genome," *New York Times*, April 27, 2002, A1.

6.   Venter told the story of his NIH days in *Life Decoded*, 89–158.

7.   Venter, *Life Decoded*, 101; Shreeve, *Genome War*, 78.

8.   Watson described the PCR process in *DNA*, 174–76. For a technical discussion of the PCR process, see James D. Watson et al., *Recombinant DNA*, 2nd ed. (New York: Scientific American Books, 1992), 79–82.

9.   Venter, *Life Decoded*, 100; Watson, *DNA*, 177. For a technical discussion of changes in DNA sequencing machines, see Watson et al., *Recombinant DNA*, 3rd ed., 259–63.

10. Mark D. Adams et al., "Complementary DNA Sequencing: Expressed Sequence Tags and Human Genome Project," *Science* 252, no. 5013 (June 21, 1991): 1651–56. See also Venter, *Life Decoded*, 95–96, 120–22, 126–27, 168; Shreeve, *Genome War*, 81–82; Gina Kolata, "Biologist's Speedy Gene Method Scares Peers but Gains Backer," *New York Times*, July 28, 1992, C1. In contrast, in *DNA*, 180, Watson asserted that Sydney Brenner pioneered the cDNA approach and Venter learned about it on a visit to Brenner's lab.

11. Venter, *Life Decoded*, 130–33, 135–38, 150, 175–76; Shreeve, *Genome War*, 83–86; Cook-Deegan, *Gene Wars*, 311–12, 317–24.

12. For the Venter–Watson feud, see Venter, *Life Decoded*, 111–13, 116–17, 122–23, 127–28, 144–48; Shreeve, *Genome War*, 79, 82–85; Cook-Deegan,

Gene Wars, 312–17. See also, Jerry E. Bishop and Hilary Stout, "Gene Scientist to Leave NIH, Form Institute," *Wall Street Journal*, July 7, 1992, B1; Robin Herman, "NIH Genes Researcher is Leaving for his Own Lab," *Washington Post*, July 7, 1992, Health Supplement, 4, Kolata, "Biologist's Speedy Gene Method"; Christopher Anderson, "Controversial NIH Genome Researcher Leaves for New $70–Million Institute," *Nature* 358, no. 6382 (July 9, 1992): 95.

13. Robert Cook-Deegan, "The Colussus of Codes," in *Inspiring Science: Jim Watson and the Age of DNA*, ed. John R. Inglis, Joseph Sambrook, and Jan A. Witkowski (Cold Spring Harbor, NY: Cold Spring Harbor Laboratory Press, 2003), 393.

14. The Institute for Genomic Research, 2004 Internal Revenue Service Form 990, Part III.

15. Venter, *Life Decoded*, 153–55, 158–60, 162–65; Shreeve, *Genome War*, 86–87. See also Kolata, "Biologist's Speedy Gene Method"; Bishop and Stout, "Gene Scientist to Leave NIH"; Herman, "NIH Genes Researcher"; Lawrence M. Fisher, "Mining the Genome," *New York Times*, January 30, 1994, A1; John Carey et al., "The Gene Kings," *Business Week* 3423 (May 8, 1995): 72–78; Beth Berselli, "Gene Split," *Washington Post*, July 7, 1997, F5.

16. Shreeve, *Genome War*, 86.

17. Human Genome Sciences, Inc. (HGS), Securities and Exchange Commission (SEC) Form 10-K, March 31, 1997, Collaborative Arrangements, n.p.; HGS, SEC Form 10-K, March 31, 1998, 10–11.

18. Venter, *Life Decoded*, 164–65. According to Shreeve, *Genome War*, 87–88, HGS invoked the extension clause on any sequence that had any possible medical importance. In contrast, William Haseltine, CEO of HGS, asserted that HGS had a policy of permitting publications as soon a patent was filed on any genes TIGR discovered and in all requested cases, before the six-month period expired. Berselli, "Gene Split." In early 1997, HGS disclosed: "[HGS] and TIGR have had recent disagreements concerning the scope of [TIGR's] non-disclosure obligations. It has come to [HGS'] attention that certain disclosures by TIGR of sequence and other information which [HGS] believes may violate such non-disclosure obligations may have taken place or may take place in the future. Disclosure of information by TIGR in violation of its non-disclosure obligations may negatively affect [HGS'] ability to obtain patent protection on the inventions described therein." HGS, SEC Form 10-K, March 31, 1997, Collaborative Arragements, n.p.

19. SmithKline Beecham, Press Release, "SmithKline Beecham and Human Genome Sciences Announce Collaboration For Large Scale Gene Sequencing," May 20, 1993; HGS SEC Form 10-K, March 31, 1997, Collaborative Arrangements, n.p. and Note E to Financial Statements, F-9; HGS, SEC Form 10-K, March 31, 1998, 8, and Note E to Financial Statements, F-9. See also Venter, *Life Decoded*, 172; "SmithKline Pact with HGS," *Wall Street Journal*, May 21, 1993, http://ProQuest (accessed December 1, 2009); Stephen D. Moore, "SmithKline Maps Out Research Route — Drug Maker on New Road With Gene Data Company," *Wall Street Journal*, September 7, 1993, http://ProQuest (accessed December 1, 2009). For a retrospective

look at the benefits of the HGS/SmithKline collaboration, see Kathleen Day, "It's All in the Genes," *Washington Post*, September 30, 1996, F14.

20. HGS, SEC Form 10-K, March 31, 1997, Collaborative Arragements, n.p., and Note E to Financial Statements, F-10; HGS, SEC Form 10-K, March 31, 1998, 8–9 and Note E to Financial Statements, F-9; HGS, Press Release, "Human Genome Sciences and SmithKline Beecham Sign Revised Genomics Collaboration Agreement," July 2, 1996. See also Lawrence M. Fisher, "Drug Makers Set to License Data on Genes," *New York Times*, July 3, 1996, D3; Kathleen Day, "Human Genome Gains Major Research Funds," *Washington Post*, July 3, 1996, F1; "Human Genome Sets More Collaborations in Gene Development," *Wall Street Journal*, July 3, 1996, B6; "Human Genome Says It Plans to License Database to Merck," *Wall Street Journal*, April 19, 1996, B6; "Genome Technology Agreement," *Wall Street Journal*, July 17, 1996, B4; "Merck of Germany to Explore Genetics Technology," *New York Times*, July 17, 1996, D3. HGS also entered into the following collaboration agreements: June 1995, Option and License Agreement with Takeda Chemical Industries, Ltd., receiving $5 million; January 1996, Research and Collaboration Agreement with Pioneer Hi-Bred International, Inc., receiving $16 million over three years; March 1996, License Agreement with F. Hoffman-LaRoche, Ltd., receiving $2 million; June 1996, Collaboration and License Agreement with Schering-Plough, receiving $5 million in fees over five years; October 1996 License and Research Agreement with Pharmacia & Upjohn Co., receiving $3 million. HGS, SEC Form 10-K, March 31, 1997, Collaborative Arrangements, n.p and note E to Financial Statements, F-11 to F-12 and HGS, SEC Form 10-K, March 31, 1998, 8–10 and note E to Financial Statements, F-9 to F-11.

21. Venter, *Life Decoded*, 180–82, 203–4, 218.

22. Quoted in Berselli, "Gene Split."

23. HGS, SEC Form 10-K, March 31, 1997, Collaborative Arrangements, n.p; HGS, SEC Form 10-K, March 31, 1998, 11. See also Venter, *Life Decoded*, 181. After TIGR, in October 1996, notified the other two parties of its decision to terminate the agreement, the document, according to its terms, terminated in April 1997. HGS then removed its sequences from the database; the remaining TIGR sequences were made publicly available without restriction.

24. Jerry E. Bishop, "Pioneer Scientist in Genetic Research Says He'll Soon Make Public Secret Data," *Wall Street Journal*, September 30, 1994, B5; Elliott Marshall, "A Showdown over Gene Fragments," *Science* 266, no. 5183 (October 14, 1994): 208–9.

25. Venter, *Life Decoded*, 210. See also Elyse Tanouye, "Gene Pioneer Opens His Databank," *Wall Street Journal*, September 28, 1995, B1; David Brown and Rick Weiss, "Scientist Glimpse Genes' Division of Labor," *Washington Post*, September 28, 1995, A3.

26. Mark D. Adams et al., "Initial Assessment of Human Gene Diversity and Expression Patterns Based upon 83 Million Nucleotides of cDNA Sequence," *Nature* 377, no. 6547 Supplement (September 28, 1995): 3–174.

27. Venter described the shotgun method in *Life Decoded*, 107, 119, 191, 196–97, 202, 246, 261–63, 288–89; J. Craig Venter et al., "Shotgun Sequencing of the Human Genome," *Science* 280, no. 5369 (June 5, 1998): 1540–42; Prepared Statement of J. Craig Venter, Ph.D., President and Chief Scientific Officer, Celera Genomics, A PE Corporation Business, Before the Subcommittee on Energy and Environment, Committee on Science, U.S. House of Representatives, 106th Congress, 2nd Session, April 6, 2000. See also Douglas Birch, "Seduction of a Scientist," *Baltimore Sun*, April 12, 1999, 1A; James L. Weber and Eugene W. Myers, "Human Whole-Genome Shotgun Sequencing," *Genome Research* 7, no. 5 (May 1, 1997): 401–9.

28. For a technical discussion of the use of computer algorithms and powerful computers, see Watson, *Recombinant DNA*, 3rd ed., 263–64, 266–67.

29. Robert D. Fleischmann et al., "Whole-Genome Random Sequencing and Assembly of *Haemophilus influenzae Rd*," *Science* 269, no. 5233 (July 28, 1995): 496–512; Hamilton O. Smith et al., "Frequency and Distribution of DNA Uptake Signal Sequences in the *Haemophilus influenzae Rd* Genome," *Science* 269, no. 5223 (July 28, 1005): 538–40. See also Venter, *Life Decoded*, 193–203, 204–5, 207; Nicholas Wade, "Bacterium's Full Gene Makeup is Decoded," *New York Times*, May 26, 1995, A16; Nicholas Wade, "First Sequencing of Cell's DNA Defines Basis of Life," *New York Times*, August 1, 1995, C1; Nicholas Wade, "Thinking Small Paying Off Big in Gene Quest," *New York Times*, February 3, 1997, A1; Kathleen Day, "Local Lab Helps Crack a Germ's Genetic Code," *Washington Post*, July 28, 1995, C1.

30. Claire M. Fraser, "The Minimal Gene Complement of *Mycoplasma genitalium*," *Science* 270, no. 5235 (October 20, 1995): 397–404. See also Venter, *Life Decoded*, 208; Karen Y. Kreeger, "First Completed Microbial Genomes Signal Birth of New Area of Study," *The Scientist* 9, no. 23 (November 27, 1995): 14–15; Andre Goffeau, "Life with 482 Genes," *Science* 270, no. 5235 (October 20, 1995): 445–46.

31. Quoted in Nicholas Wade, "Genome Project Partners Go Their Separate Ways," *New York Times*, June 24, 1997, C2. For background on the Venter–Haseltine feud with respect to the *H. flu* genome, see Venter, *Life Decoded*, 203–4; Shreeve, *Genome War*, 107–9.

32. HGS, SEC Form 10-K, March 31, 1998, 2, 11 and Note D to Financial Statement F-9. See also Venter, *Life Decoded*, 216–18, 220–21, 226; Berselli, "Gene Split"; Wade, "Genome Project Partners"; Beth Berselli, "Genetic Research Firm Severs Ties with Nonprofit," *Washington Post*, June 24, 1997, C4; Evelyn Strauss, "Corporate Divorce Reveals Genetic Secrets," *Science News*, 152, no. 2 (July 12, 1997): 29.

33. Venter, *Life Decoded*, 220.

34. I have drawn on Venter, *Life Decoded*, 227–53; Shreeve, *Genome War*, 113, 117–18; "Celera is Name of New Genomics Company Formed by Perkin-Elmer and Dr. J. Craig Venter," *Business Wire*, August 5, 1998. See also Nicholas Wade, "Scientists Plan: Map All DNA Within 3 Years," *New York Times*, May 10, 1998, A1; Bill Richards, "Perkin-Elmer Jumps into Race to Decode Genes," *Wall Street Journal*, May 11, 1998, B6; Bill Richards, "Perkin-Elmer Will Join Venture to Decode Genes," *Wall Street Journal*,

May 13, 1998, CA2; Nicholas Wade, "The Genome's Combative Entrepreneur," *New York Times*, May 18, 1999, F1; Justin Gillis and Rick Weiss, "Private Firm Aims to Beat Government to Gene Map," *Washington Post*, May 12, 1998, A1; Justin Gillis, "One Man's Race to Map Genetic Code," *Washington Post*, August 22, 1998, A1. In *Genome War*, 65–67, Shreeve detailed how the idea for decoding the human genome as a private venture germinated at a meeting of Perkin-Elmer executives in November 1997.

35. Shreeve, *Genome War*, 113.
36. Perkin-Elmer Corp. (Perkin-Elmer), Press Release, "Perkin-Elmer, Dr. J. Craig Venter, and TIGR Announce Formation of New Genomics Company," May 9, 1998; Perkin-Elmer, SEC Form 10-K, September 25, 1998, 1, 9.
37. Quoted in Perkin-Elmer, Press Release, "Perkin-Elmer, Dr. J. Craig Venter."
38. Venter, *Life Decoded*, 236.
39. Ibid., 233, 235, 237, 265; Shreeve, *Genome War*, 117.
40. J. Craig Venter, Prepared Statement before the Subcommittee on Energy and Environment.
41. PE Corp., SEC Form 10-K, September 28, 1999, 13–14. See also PE Corp., SEC Form 10-K September 28, 2000, 13.
42. Perkin-Elmer, Press Release, "Perkin-Elmer's Celera Genomics to Provide Amgen Access to New Database Products," January 12, 1999; Celera, Press Release, "Celera Genomics and Pharmacia & Upjohn Enter Into Five-Year Subscription Agreement for Celera Database Products," March 17, 1999; Celera, "Celera Genomics Enters into Five-year Database Agreement with Novartis," April 19, 1999; Celera, Press Release, "Celera Genomics to Provide Target Gene Discovery Services and Human, Mouse and SNP Databases to Pfizer," November 22, 1999.
43. Shreeve, *Genome War*, 163. Shreeve detailed other aspects of the Amgen agreement. Ibid.
44. EG&G, Inc., Press Release, "EG&G to Acquire Perkin-Elmer's Analytical Instruments Division for $425 Million," March 8, 1999. See also "EG&G Plans to Buy Unit of Perkin-Elmer In $425 Million Deal," *Wall Street Journal*, March 9, 1999, B8.
45. Shreeve, *Genome War*, 317.
46. Joint Statement by President Clinton and Prime Minister Tony Blair of the United Kingdom on the Availability of Human Genome Data, March 14, 2000 in *Public Papers of the Presidents of the United States: William J. Clinton, 2000–2001*, Book 1—January 1 to June 26, 2000 (Washington, DC: U.S. Government Printing Office, 2001), 462. See also Alex Berenson and Nicholas Wade, "A Call for Sharing of Research Causes Gene Stocks to Plunge," *New York Times*, March 15, 2000, A1; Robert Langreth and Bob Davis, "Press Briefing Set Off Rout in Biotech," *Wall Street Journal*, March 16, 2000, A19.
47. Venter, *Life Decoded*, 267–80; Shreeve, *Genome War*, 255–63, 268–78, 281–82.
48. Mark D. Adams et al., "The Genome Sequence of *Drosophila melanogaster*," *Science* 287, no. 5461 (March 24, 2000): 2185–95; Gerald M. Rubin

et al., "Comparative Genomics of the Eukaryotes," *Science* 287, no. 5461 (March 24, 2000): 2204–15; Eugene W. Myers, "A Whole-Genome Assembly of *Drosophila*," *Science* 287, no. 5461 (March 24, 2000): 2196–204. See generally Michael Ashburner, *Won for All: How the Drosophila Genome Was Sequenced* (Cold Spring Harbor, NY: Cold Spring Harbor Laboratory Press, 2006), 1–48. See also Venter, *Life Decoded*, 244, 267–69, 271–79; Nicholas Wade, "On Road to Human Genome, A Milestone in the Fruit Fly," *New York Times*, March 24, 2000, A1; Justin Gillis, "Mapping of Fruit Fly a Genetic 'Milestone,'" *Washington Post*, March 24, 2000, A1; Robert Langreth, "PE's Celera Genomics Decodes Entire Genome of the Fruit Fly," *Wall Street Journal*, March 24, 2000, B6; "Celera Genomics Says Fruit-Fly Sequencing is Mostly Completed," *Wall Street Journal*, September 10, 1999, B8.

49.   Venter, *Life Decoded*, 276; Shreeve, *Genome War*, 299–300.

50.   Venter, *Life Decoded*, 286.

51.   Ibid., 269–71, 292.

52.   Ibid., 265–66.

53.   Applera Corp. (Applera), Press Release, "J. Craig Venter Steps Down as President of Celera Genomics," January 22, 2002; Venter, *Life Decoded*, 259, 327–28. See also Shreeve, *Genome War*, 369–70; Andrew Pollack, "Scientist Quits The Company He Led in Quest for Genome," *New York Times*, January 23, 2002, C1; Scott Hensley, "Venter Leaves Celera as Science, Business Clash," *Wall Street Journal*, January 23, 2002, B1; Justin Gillis, "New Chief of Celera to Weigh Overhaul," *Washington Post*, January 24, 2002, E1; Justin Gillis, "Celera Changed, Venter Couldn't," *Washington Post*, January 28, 2002, E1; Andrew Pollack, "The Genome is Mapped. Now He Wants Profit," *New York Times*, February 24, 2002, Business Section 1.

54.   Venter, *Life Decoded*, 259. On Celera's original business plan, see Ibid., 292–93. See also Shreeve, *Genome War*, 120–22, 338–39; Justin Gillis, "The Gene Map and Celera's Detour," *Washington Post*, February 21, 2001, E1; Scott Hensley, "Celera Discloses Data Rates Academic Researchers Pay," *Wall Street Journal*, July 11, 2001, B8.

55.   Celera, Press Release, "Celera Completes Assembly of Mouse Genome," April 27, 2001. See also Justin Gillis, "Celera Has Mouse Map Monopoly," *Washington Post*, April 27, 2001, E1; Nicholas Wade, "Genetic Sequence of Mouse is Also Decoded," *New York Times*, February 13, 2001, F5; Scott Hensley, "Progress on Deciphering Genetics of Mice Comes Surprisingly Fast," *The Wall Street Journal*, February 16, 2001, B2; Scott Hensley, "Celera Completes Mouse Genome," *Wall Street Journal*, April 27, 2001, B5.

56.   Andrew Pollack, "Celera to Quit Selling Genome Information," *New York Times*, April 27, 2005, C2.

57.   Applera, SEC Form 10-K, September 26, 2001, 21–22; Applera, SEC Form 10-K, September 27, 2002, 28–29; Celera Diagnostics Joint Venture Agreement, as of April 1, 2001 in Applera, SEC Form 10-K, September 27, 2002, Exhibit 10.36.

58.   Applera, Press Release, "Applera Corporation's Three Businesses—Applied Biosystems, Celera Genomics and Celera Diagnostics—in Comprehensive

Program to Commercialize Products Based on Discoveries from Human Genome," July 24, 2001; Applera, SEC Form 10-K, September 26, 2001, 15; Applera, SEC Form 10-K, September 27, 2002, 3, 19–20. See also Terence Chea, "Celera Aims to Spot Genetic Differences," *Washington Post*, July 25, 2001, E5; Andrew Pollack, "Race Under Way to Winnow Down Genetic Data," *New York Times*, July 24, 2001, C1; Scott Hensley, "Applera to Catalog Genetic Variations," *Wall Street Journal*, July 24, 2001, B6.

59. Celera, Press Release, "Celera Genomics to Acquire Axys Pharmaceuticals, Inc.," June 13, 2001; Applera, SEC Form 10-K, September 26, 2001, 17–18; Applera, SEC Form 10-K September 27, 2002, 2, 22–23. See also Justin Gillis, "Celera Buys into Drug Research," *Washington Post*, June 14, 2001, E1; Scott Hensley, "Celera to Buy Axys for $174 Million," *Wall Street Journal*, June 14, 2001, B3; Andrew Pollack, "Genome Research Pioneer to Buy Drug Maker," *New York Times*, June 14, 2001, C10.

60. Samuel P. Dickson et al., "Rare Variants Create Synthetic Genome-Wide Associations," *PLoS Biology* 8, no. 1 (January 2010): e1000294. See also Nicholas Wade, "A New Way to Look for Diseases' Genetic Roots," *New York Times*, January 26, 2010, D4.

61. See generally, Francis S. Collins, *The Language of Life: DNA and the Revolution in Personalized Medicine* (New York: HarperCollins, 2010). For one case, possibly the first in which the results of DNA sequencing altered a patient's treatment, see Murim Choi et al., "Genetic Diagnosis by Whole Exome Capture and Massively Parallel DNA Sequencing," *Proceedings of the National Academies of Sciences* 106, no. 45 (November 10, 2009): 19096–101. See also Matthew Herper, "A First: Diagnosis by DNA," *Forbes* 185, no. 3 (March 15, 2010): 56. For a background on whole-genome sequencing tests, see Amy Dockser Marcus, "How Genetic Testing may Spot Disease Risk," *Wall Street Journal*, May 4, 2010, D3; Andrew Pollack, "Outlook Uncertain," *New York Times*, March 20, 2010, B1; Nicholas Wade, "Disease Cause is Pinpointed with Genome," *New York Times*, March 11, 2010, A1; Jared C. Roach et al., "Analysis of Genetic Inheritance in a Family Quartet by Whole-Genome Sequencing," *Science* 328, no. 5978 (April 30, 2010): 636–39.

62. Jeanne Whalen, "DeCODE, Pioneer in Genome Studies, Seeks Bankruptcy Protection," *Wall Street Journal*, November 18, 2009, B4; Nicholas Wade, "A Genetics Company Fails, its Research too Complex," *New York Times*, November 18, 2009, B2.

63. Ron Winslow and Jennifer Corbett Dooren, "Gene Work Yields New Treatment for Lupus," *Wall Street Journal*, March 10, 2011, A1; Andrew Pollack, "F.D.A. Approves Drug for Lupus, an Innovation after 50 Years," *New York Times*, March 10, 2011, B1. See also HGS, Press Release, "Human Genome Science and GlaxoSmithKline Announce Positive Results In Second of Two Phase 3 Trials of Benlysta™ In Systemic Lupus Erythematosus," July 20, 2009. See also Peter Benesh, "Glitches Hinder OKs for Human Genome," *Investor's Business Daily*, April 26, 2010, A8; Andrew Pollack, "In Trials, A New Drug for Lupus Advances," *New York Times*, November 3, 2009, B1; Mike Musgrove, "Lupus Drug Headed to FDA," *Washington*

Post, November 3, 2009, A13; Thomas Gryta, "Experimental Lupus Drug Shows Promise in New Study," *Wall Street Journal*, November 3, 2009, B6; Rebecca Ruiz, "The Lupus Drug Gold Mine," *Forbes* 185, no. 1 (February 8, 2010): 40.

64.   Venter, *Life Decoded*, 331.

Part III
The Quest to Write Life

5

Organizational Structure and Funding of Nonprofit Entities

The human genome race made J. Craig Venter rich and famous. After being ousted from Celera in January 2002, devastated, he fell into a deep funk. Turning to retail therapy, Venter bought a $5 million villa high on a mountainside overlooking the Caribbean Sea.[1] The good life left him restless, however, and in several months, the old J. Craig Venter resurfaced, rejuvenated, and ready for the future.

Venter embarked on an ambitious new quest, transforming his purpose from reading the genetic code to writing it. His ultimate goal: to create a new genome from scratch. In creating synthetic life, he sought to assemble a functioning organism, a new biological entity from scratch, using only necessary genes, while providing the foundation for new energy and petrochemical industries along the way. It was an ambitious goal. Synthetic organisms could produce biofuels, pharmaceuticals, and substitutes for petrochemical products, such as plastics.

In contrast to his dealings in the 1990s with Steinberg and Perkin-Elmer (discussed in Chapter 5), after his ousting at Celera Genomics Corp., Venter wanted to be master of his own destiny. Having more than $150 million from his Celera and HGS stock to work with, Venter could now do the science he wanted in the way he wanted to do it.[2] In his quest to develop disruptive technologies[3] based on synthesizing and constructing whole chromosomes and genomes, he focused on one thing: using living systems to promote sustainability and to increase the likelihood of humanity's survival as a species, for example, by creating renewable fuels to replace petroleum and coal. Venter sought to apply his expertise to benefit society, noting, "This is a unique opportunity to see if our science can literally help save the planet."[4]

This chapter discusses Venter's organizational structure and the funding of his nonprofit entities post-Celera. He again implemented a two-part organizational structure, putting together academically oriented nonprofit entities to conduct scientific research and policy activities and a for-profit unit to engage in applied research and translate scientific breakthroughs into viable commercial solutions. Thereafter, Chapter 6 provides an overview of the research progress in synthetic biology made by Venter's nonprofit team. Venter's commercialization efforts are examined in Chapter 7, and his bioenergy competitors are discussed in Chapter 8.

Formation of Nonprofit Entities

On the nonprofit side, Venter established multiple entities in quick succession, starting in 2002, in addition to the already existing TIGR, founded in July 1992. Even with the new nonprofits on the scene, TIGR continued to function as one of the world's leading centers for researching microbial, parasitic, and eukaryotic genomes.

In 1998, Venter's then-wife, Claire M. Fraser, Ph.D., had taken over as President of TIGR when Venter left for Celera. Since then, Fraser had built TIGR's staff to three hundred people, with some $40 million in annual research grants. But when Venter was ousted from Celera and established the additional nonprofit entities, he did not return to his former role at TIGR. "So I said it was much better, rather than disrupting that structure, to form these sister organizations where I could play a role," Venter noted.[5] Despite his hands off approach to TIGR and his involvement in the other organizations, Venter continued in his previous role at TIGR as a financial pillar, having already endowed TIGR with about $30 million in HGS shares, and as Chairman of the Board of Trustees.[6]

After being fired as CEO of Celera and taking some time off, in the spring of 2002 Venter got started again founding four separate nonprofit organizations: J. Craig Venter Science Foundation, Inc.; The Center for the Advancement of Genomics (TCAG); Institute for Biological Energy Alternatives (IBEA); and the J. Craig Venter Science Foundation Joint Technology Center, Inc (JTC). He served as president and chairman of each of these separate but interrelated entities.[7]

J. Craig Venter Science Foundation, Inc.

In April 2002, Venter unveiled the J. Craig Venter Science Foundation, Inc. (Science Foundation). Funded by his Celera stock holdings, the Science Foundation served mostly as a financial support organization for the benefit of not only TIGR, but also for three other of his nonprofits: TCAG; IBEA; and JTC. The Science Foundation also provided administration support, coordinated policy and research activities among TCAG, IBEA, and TIGR, provided investment management and fund-raising on behalf of these three organizations, and explored new ways to foster science education and scientific innovation.[8]

The Center for the Advancement of Genomics

At the same time as Venter established his foundation, he set up a new nonprofit institute, TCAG, funded by the Science Foundation. TCAG served as a public policy center focusing on helping the general public and elected officials better understand the social and ethical implications of genomics research, including genetic nondiscrimination and the use of stem cells.[9] Its scientists also explored the relationship between the genetic code and human traits and behavior.

Institute for Biological Energy Alternatives, Inc.

Venter also started the IBEA.[10] Looking for new organisms and analyzing known organisms that metabolize carbon, as scientific research institution IBEA sought ways to use biology and genetics, including microbes, microbial genomics, and microbial pathways, to produce "clean" fuels and reduce the amount of carbon dioxide released in the atmosphere from petroleum and coal. IBEA's efforts were devoted mainly to research focused on "exploring solutions for carbon sequestration using microbes, microbial pathways, and plants."[11] To achieve this goal, IBEA sought to apply genomics to enhance the ability of terrestrial and oceanic microbial communities to remove carbon from the atmosphere and thus lessen global warming. IBEA also studied how microorganisms digest raw materials so that it could develop modified organisms that used new biological pathways to produce nonpolluting fuels such as hydrogen, then viewed as a promising alternative to petroleum-based energy sources.

Venter also developed a vision for a new field of environmental genomics. The shotgun sequencing of the life in oceans, as he noted, "could provide a snapshot of ocean health today, help to monitor its health tomorrow, and reveal the microbes responsible for creating much of our atmosphere."[12] Venter also hoped that the study of the metabolic machinery of ocean microbes might also lead to new ways to make alternative fuel sources.

J. Craig Venter Science Foundation Joint Technology Center, Inc.

Carrying out Venter's vision of environmental genomics required a substantial DNA sequencing facility. In January 2003, he established another new nonprofit organization, the JTC to do the sequencing for TIGR as well as IBEA.[13] He funded the new center with nearly $28 million.[14] TIGR, TCAG, and IBEA began building the joint technology facility to rapidly and accurately sequence and analyze genomes in a more cost-effective manner. In announcing the formation of the JTC, Venter stated, "Our goal is to build a new and unique sequencing facility that can deal with the large number of organisms to be sequenced, and can further analyze those genomes already completed.... The evolution of sequencing has advanced greatly in the last decade. My teams at TIGR and later at Celera took large-scale sequencing to new heights, and the new center will be the latest in the evolution of these facilities."[15] The facility used the most up-to-date automated sequencing, supercomputing, networking, and high-performance storage technologies and allowed Venter's nonprofit organizations to expand and speed up the pace of research on a wide array of projects. The facility greatly increased the organizations' sequencing capacity and gave researchers the tools to tackle large genomes more rapidly and at a lower cost.

Subsequent Nonprofit Organizational Developments

After the formation of the four entities, Venter's nonprofit structure went through three reorganizations in 2004, 2006, and 2007. In September 2004, Venter consolidated three of his research institutes, TCAG, IBEA, and JTC, into one institute, the Venter Institute.[16] At that time, instead of being merged into the new institute, TIGR remained as an affiliated research organization, and the Science Foundation continued as a supporting organization, funding both TIGR and the Venter Institute while also providing administrative support, coordinating policy and research activities, and carrying out investment management and fund-raising on behalf of these two organizations. By merging the three institutes (TCAG, IBEA, and JTC) into one, Venter sought to streamline administrative, board, and fiscal functions so that the combined institute could more effectively concentrate on research tasks. The Venter Institute continued the groundbreaking research begun at TCAG, IBEA, and JTC in several major areas including: human genomic medicine; environmental and evolutionary genomics; synthetic biology; biological energy production; and high throughput DNA sequencing.

Then, in October 2006, Venter merged TIGR and the Science Foundation into the Venter Institute.[17] At that time, the research organization formerly known as the J. Craig Venter Institute, previously formed in 2004, was renamed TCAG. Venter served as Chairman and Chief Executive Officer of the new Venter Institute. TIGR, now a division of the Venter Institute, continued to be led by Venter's former wife, Claire Fraser-Liggett, as President. TCAG (the old J. Craig Venter Institute), now a division of the new Venter Institute, was led by a new president, Robert Strausberg, Ph.D. Consolidating the three affiliated organizations into one nonprofit research institute represented a prudent step financially, administratively, and scientifically. Building on TIGR and Venter Institute's genomic breakthroughs, the unified organization, it was hoped, would be an even greater force in genome research.

Six months later, in April 2007, the Venter Institute revamped its organizational structure.[18] At that time, according to Venter, "Since the earliest days of founding TIGR in 1992 and then with the other affiliated institutes, my goal has always been to create unique and dynamic research organizations that push the boundaries of traditional science. We have long been leaders in genomics and with our newly organized Institute, I am certain we are poised to continue to blaze new trails in the field."[19] Instead of being organized into two research divisions, as implemented in 2006, it now consisted of an administrative team and ten research groups including: Genomic Medicine; Infectious Diseases; Synthetic Biology & Bioenergy; Plant Genomics; Microbial & Environmental Genomics; Pathogen Functional Genomics Resource Center; Applied Bioinformatics; Research Informatics; Software Engineering; and the Policy Center. Even after reorganization, genomic sequencing capability remained a cornerstone of the Venter Institute's activities, and Venter continued as the institute's Chairman and President, with Robert Strausberg, serving as deputy director.

Today, the Venter Institute is one of the world's largest private research institutes with more than five hundred scientists and staff (including nearly four hundred dedicated to research), more than $200 million in assets, and an annual budget of more than $86 million.[20] The West Coast Venter Institute, a new arm of Venter's research empire located in La Jolla, California, opened in 2008. The institute, a world leader in genomic research, as a focused multidisciplinary genomic organization, presently concentrates its efforts on three fields: Genomic Medicine; Environmental Genomics; Synthetic Biology and Bioenergy.

The Venter Institute's genomic medicine researchers explore the genetic causes and genomic solutions to diseases, such as cancer, and diseases of the heart, lungs, and blood.[21] The group has ongoing collaborations with notable researchers examining physical, more technically, somatic, changes in genetic information in tumor cells in comparison with healthy cells. Because 95–97 percent of cancers are caused by somatic changes in our genetic code, rather than by genes inherited from a parent, the Genomic Medicine group's goal is not the early detection of cancer, but rather more effective treatment of the cancer by targeting the exact defect which initially caused it.[22] The group also works on similar projects with other diseases.

The Environmental Genomics group focuses primarily on the Global Ocean Sampling Expedition and other research expeditions, carried out aboard Venter's Sorcerer II yacht. A Venter limited liability company (LLC) owns the Sorcerer II, a 95-foot yacht that Venter purchased in 2000,[23] and the Venter Institute charters the yacht from the Venter LLC for oceanographic research. Selecting the yacht for charter due to its availability, its environmental friendliness, and its relatively low cost compared to other vessels, the institute paid Venter's LLC one dollar and assumed responsibility for insurance, maintenance, and operating expenses during the term of the charter.[24]

Since 2003, the Sorcerer II has been collecting biological samples from various waters around the globe and, in 2006, the Global Ocean Sampling Expedition culminated in the Sorcerer II's complete circumnavigation of the earth, during which the yacht sampled sea water approximately every two hundred miles. More recently, the Sorcerer II gathered samples from the Black, Baltic, and Mediterranean Seas as part of the J. Robert Beyster and Life Technologies 2009–2010 Research Voyage.

The Sorcerer II collects various sized microorganisms by filtering seawater samples through decreasing sized filters onboard the ship. Back in the institute's laboratories, the DNA of the microorganisms is isolated, sequenced, and analyzed, producing a massive amount of new information about oceanic life. For instance, the Global Ocean Sampling Expedition ultimately uncovered more than six million new genes.[25]

In the area of Synthetic Biology and Bioenergy, the institute sought to use its pioneering genomic science to explore and produce biologically driven energy sources. Its researchers focus on developing synthetic organisms able to produce renewable fuels and various petrochemical products.[26] In order to do so, the Synthetic Biology and Bioenergy group's goal is to create a catalog of useful genes from known organisms and use those genes to create a "superproductive organism" by altering present-day species or even creating a new species from scratch. In 2010, the team took an important step toward realizing its ultimate goal by synthesizing a bacterial genome from chemical compounds, inserting the gene into an empty cell, and using the synthesized genome to "boot up" the cell, creating the first cell controlled by a totally synthetic genome.[27] These research successes are considered in more detail in Chapter 6.

External Funding

Apart from Venter's self-funding of TIGR and the Science Foundation, his success in sequencing the human genome facilitated the flow of federal grant funds to his nonprofits from the DoE, Office of Science, National Institutes of Allergy and Infectious Diseases (part of the NIH), and the National Human Genome Research Institute (NHGRI), among other NIH units. Venter nonprofits also received funding from other tax-exempt organizations and his for-profit arm, SSGI. Beyond the scope of this book, Venter nonprofits received funding from the Naval Medical Research Center, the Defense Threat Reduction Agency, and the National Institute of Food and Agriculture, part of the U.S. Department of Agriculture (USDA).

Bioenergy Research Funding

Venter began funding his post-Celera work with some $12 million in multiyear grants from the DoE's Genomes to Life Program. These grants enabled Venter's team to explore using knowledge gained from genomics for potential alternative energy production.

The Genomes to Life Program, managed by the Energy Department's Office of Science, sought to use the department's computational capabilities and research facilities to understand the activities of single-cell organisms on three levels: first, the proteins and multimolecular machines that perform most of the cell's work; second, the gene regulatory networks that control these processes; and third, the microbial associations or communities in which groups of different microbes carry out fundamental functions in nature. By understanding how life functions at the microbial level, the program hoped that in the future its grantees would be able to use the capabilities of these organisms to help meet the nation's energy and environmental challenges.[28]

In November 2002, the Genomes to Life Program provided $3 million in funding to Venter's IBEA.[29] The three-year grant funded research to develop a synthetic chromosome as "the first step toward making a self-replicating organism with a completely artificial genome."[30] In developing cost-effective and efficient biological alternative energy sources, Venter noted, "IBEA was founded with the goal of exploring biological mechanisms for dealing with carbon sequestration and to study the creation of other potential energy sources such as hydrogen. We believe that building a synthetic chromosome is an important step toward realizing these goals because we could potentially engineer an organism with the ideal qualities to begin to cope with our energy issues."[31] With the receipt of this grant, IBEA named Hamilton O. Smith as its scientific director.

Then, in April 2003, the Department of Energy's Office of Science increased its funding of IBEA by an additional $3 million per year for the next three years.[32] Through this grant, the department sought to fund a better understanding of microbes and the development of biological methods to capture carbon dioxide from the atmosphere and produce hydrogen. With the new funds, IBEA scientists also sought to determine, using the Sorcerer II, the genetic sequences of all the microorganisms occurring in a natural microbial community, the Sargasso Sea, an environment with a manageable number of microbes. As part of the Venter Institute's environmental genomics program, the studies in the Sargasso Sea would help scientists discover previously unknown biochemical pathways and organisms that could lead to the development of new methods for carbon sequestration or alternative energy sources.

The multiyear grants totaling $12 million from the Department of Energy expired in 2005, however, and were not renewed. By 2005, Venter had already cultivated other funding sources for his non-profits, but with the loss of the Department of Energy funds, Venter turned his focus to the NIH and other tax-exempt organizations as a source of money for genomic medicine and environmental genomics research.

Genomic Medicine Funding

Grants for genomic medicine research flowed from three NIH sources: National Institute of Allergy and Infectious Diseases (NIAID); NHGRI; and Human Microbiome Project. In October 2003, TIGR signed a five-year $65 million contract with NIAID, part of the NIH, to sequence and analyze the genomes of pathogenic microbes and invertebrate transmitters, technically vectors, of infectious diseases for the wider, international scientific community.[33]

Under this contract, TIGR sequenced and analyzed genomes of eukaryotic and prokaryotic cells as well as viruses. The TIGR affiliated sequencing facility, the J. Craig Venter Science Foundation JTC, conducted the sequencing under the contract. TIGR investigators led each genome project under the grant and coordinated the resulting analysis. The contract allowed TIGR to continue to meet other researchers' needs to find new ways to prevent and treat diseases caused by pathogenic microbes.

Subsequently, in May 2009, the Venter Institute received a new $43 million, five-year contract from the NIAID.[34] The contract enabled the institute to continue to expand its expertise in infectious diseases and human genomics and provide genomic services to the broader scientific community so as to further the understanding of the microbial world and how its affects humans. Under the two NIAID contracts, initially TIGR and then the Venter Institute provided these services to the NIAID Microbial Genome Centers Program. Venter stated, "Since our first sequencing of the *Haemophilus influenzae* genome in 1995, to our most recent work in sequencing the isolates from the 2009 H1N1 [swine] flu outbreak, JCVI [J. Craig Venter Institute] is committed to being a major source for leading edge genomic data and tools to further scientific understanding of the microbial world and how it affects humans."[35]

The Venter Institute worked collaboratively with the NIAID to provide genomic resources responsive to the needs of the global infectious disease community. Its investigators with scientific and technical expertise in a variety of fields, including infectious diseases, human genomics, DNA sequencing, genotyping, and bioinformatics, generated comprehensive genomic data sets that facilitated pathogen countermeasures, such as vaccines, therapeutics, diagnostics, and surveillance efforts. Studying the microorganisms that cause infectious diseases, researchers sought to develop more effective vaccines and treatments for old and new diseases, such as H1N1, as well as counter potential bioterrorist attacks from a wide variety of microorganisms including viruses, bacteria, protozoa, parasites, and fungi.

Separate from the contract with the NIAID, in November 2003, Venter's former nemesis in the race to sequence the human genome, the NHGRI, part of the NIH, selected five centers to carry out a new generation of large-scale sequencing projects designed to maximize the promise of the Human Genome Project and dramatically expand the understanding of human health and disease. The five centers, comprising the Large-Scale Sequencing Research Network, used high-throughput robotics to sequence a strategic set of animal genomes. For fiscal year 2004, NHGRI earmarked $163 million for these sequencing centers. As part of the grant, TIGR and JTC received $10 million in initial funding.[36]

The five centers operated under cooperative agreements in which substantial programmatic involvement occurred among NHGRI and the grant recipients in the performance of the scientific activities. The centers' primary mission focused on producing a publicly available resource of high-quality, assembled genome sequences that researchers could use to address human biology and health. By comparing genome sequences from chosen organisms, scientists identified specific DNA sequences conserved throughout the evolution of different species. Researchers believe that these sequences, many of which are present in humans, reflect functionally important regions of the human genome, thereby enhancing their understanding of how the human genome works. The large-scale sequencing program also sought to develop significant improvements in sequencing efficiency and cost so that sequencing would remain a cutting-edge technology for modern biology.

In December 2007, the NIH launched the Human Microbiome Project, part of the NIH Roadmap for Medical Research. The roadmap represented a series of initiatives designed to pursue major opportunities and fill gaps in biomedical research that no single NIH unit could tackle alone, but which the agency could address to make the biggest possible impact on medical research projects. The human microbiome represents the collective microorganisms present in or on the human body and their impact, good and bad, on human health. The Human Microbiome Project sought to sequence these organisms to determine their biological impact, and the Venter Institute was chosen as one center to participate in this research.

As part of a $115 million five-year effort, the NIH awarded $8.2 million in one-year awards to four sequencing centers, including the Venter Institute, to start building a framework and data resources for the project.[37] The initial work focused on sequencing the genomes of two hundred microbes that were isolated from the human body as part of a collection that would total some one thousand microbial genomes, thereby providing a resource for investigators interested in exploring the human microbiome. The project sought to discover what microbial communities exist in different parts of the human body and how these communities change in the presence of human health or disease. Researchers used advances in next-generation DNA sequencing techniques, relying on a process called metagenomic sequencing, a genetic sequencing technique that examines the interplay between a species' genetic makeup, other microbial species, and the environment in which an organism lives, in this case, the human body.

Subsequently, in May 2009, the Human Microbiome Project awarded more than $42 million in a new round of funding (bringing the total to $157 million) to support the work of the large-scale DNA sequencing centers that participated in the initial phase of the project. The Venter Institute received an additional $8.8 million spread over four years[38] to spur advances in understanding how microorganisms that live in or on our bodies impact human health or disease. The designated centers worked together to sequence at least four hundred microbial genomes, in addition to about five hundred microbial genomes already completed or in sequencing pipelines. Then, in May 2010, the Human Microbiome Project published an analysis of 178 genomes from microbes that live in or on the human body.[39] These 178 microbial genomes launched the project's reference collection that eventually will total some 900 genomes of bacteria, viruses, and fungi. Researchers will use this data to characterize the microbial communities found in samples taken from healthy human volunteers and, later, from those with specific illnesses.

Since 2002, Venter's institutes received a number of grants for genomic medicine research from the NIAID, the NHGRI, and the Human Microbiome Project that easily surpassed a total of $100 million. Venter's institutes were also focused on another field of interest, namely, environmental genomics, however, and were able to duplicate their fund-raising success in this area.

Environmental Genomics Funding

The microbial world hidden in the oceans of the earth contains a vast amount of information about how earth sustains itself. While the largest species of the oceans are the most visible to us, the aggregate effect of the oceans' cellular life has a greater impact on our lives. Since 2003, Venter has tapped his Sorcerer II yacht to sample the genomic information of these microorganisms in their native habitats.

To fund the environmental genomics work on Sorcerer II in the Sargasso Sea, Venter initially obtained funding from the DoE, Office of Science and the Science Foundation for the sequencing and analysis. With its research successes in the Sargasso Sea, as detailed in Chapter 6, the Sorcerer II Expedition then received a $4.25 million grant in March 2004 from the Gordon and Betty Moore Foundation to sequence the DNA collected along the coast of North America.[40] Six months later, in September 2004, the Venter Institute received an additional $8.9 million in a two-year grant from the Gordon and Betty Moore Foundation to support a vastly expanded reference collection of genomic analyses of marine microbes.[41] The Marine Microbe Genome Project, funded by the September 2004 grant, allowed the institute to go from less than ten completely sequenced ocean microbes to more than one hundred in one year. Data about their genetic makeup helped scientists understand the functions that microbes play in the ecology of the oceans.

Then, in January 2006, the Gordon and Betty Moore Foundation awarded a seven-year $24.5 million grant to the Community Cyber-infrastructure for Advanced Marine Microbial Ecology Research and Analysis (CAMERA) Project.[42] The project's grantees, University of California, San Diego (UCSD) computer experts, the UCSD Center for Earth Observations and Applications, the UCSD Division of the California Institute for Telecommunications and Information Technology, the Scripps Institute of Oceanography, and the Venter Institute, sought to map the genes of the tiniest marine creatures. A comprehensive picture of contemporary sea life would offer scientists new insights into how life first originated before organisms appeared on land. The project relied on metagenomics, the same technique the Venter Institute used in the Human Microbiome Project, which examined the interplay among an organism's genetic makeup, other species, and the physical characteristics of the ocean where it was found. Here the surrounding environment, the ocean, was much larger than the human body and contained a much greater body of undiscovered knowledge. According to Venter, "The explosion of data from the collection and sequencing of marine microbes requires a completely novel approach to storing, accessing, mining, analyzing, and drawing conclusions from this rich new wealth of information. The goal is to create a community resource to house all metagenomic data that will facilitate and advance knowledge of marine microbial ecology, other natural environments and evolutionary biology."[43]

The movement from traditional species-based genome databases to the CAMERA-based metagenomics data storage required increasingly complex computational abilities and the development of more sophisticated cyber-architectures. These requirements revolutionized the way genomics research is conducted. Through newly developed, advanced connectivity technologies, the CAMERA project permitted scientists from around the globe to connect local laboratory PC clusters, or groups of PCs, directly to the centralized CAMERA database and tools, resulting in up to a hundredfold increase in data transfer speeds over than existing standards. Access to all of CAMERA's computational tools as well as Venter's Global Ocean Sampling Expedition data allowed researchers to analyze information in ways that address the challenges of metagenomic analysis, specifically the ability to analyze data in innovative and comprehensive ways.

The CAMERA project, however, was not the only ongoing use that Venter found for the Sorcerer II. Funded by the Beyster Family Foundation of the San Diego Foundation, Life Technologies Corp. and anonymous donors, in March 2009, the Venter Institute announced the launch of a new Sorcerer II Expedition, the J. Robert Beyster and Life Technologies 2009–2010 Research Voyage.[44] The two-year voyage sampled microbial diversity in the Baltic, Mediterranean, and Black Seas that are among the world's largest seas isolated from major oceans. Building on the success of the earlier Sorcerer II expeditions, Venter noted, "We are confident this voyage will yield important insights into the microbial universe there and will add to the growing catalogue of microbes and genes my team has been compiling through the Sorcerer II Expedition."[45]

Funding from Venter's For-Profit Entity and a Sale-Leaseback Transaction

On the for-profit front, examined in detail in Chapter 7, in 2005, Venter cofounded Synthetic Genomics, Inc. (SGI) and launched its efforts to solve energy and environmental problems using microbes. The firm sought to develop synthetic organisms, engineered to execute specific desired functions. Specifically, SGI wanted to reprogram cells to have new, unique, and highly profitable metabolisms that serve as biofactories for alternative fuel and petrochemical sources.

The Venter Institute received funding from SGI in exchange for the exclusive assignment of the intellectual property rights to its synthetic genomics inventions. A sponsored research agreement between the institute and SGI provided substantial support for the nonprofit's synthetic biology group and funding for genomic sequencing and analysis of the genomes of oil palm and Jatropha as well as various environment genomic projects. This funding amounted to more than $9.8 million in 2007 and nearly $22.6 million in 2008.[46] SGI provided funds for these institute research projects: in 2007, the first bacterial genome transplantation changing one species to another; in 2008, the advance in genome assembly technology; in 2009, the cloning and engineering of bacterial genomes in yeast and the transplantation of these genomes back into bacterial cells; in 2010, the synthesis of a bacterial genome to take over a cell.[47] These research successes are discussed in the next chapter.

In May 2010, to raise cash the Venter Institute sold five buildings that comprised its suburban Maryland campus to BioMed Realty Trust, Inc. for $53 million.[48] The institute leased back the buildings in a ten-year renewable term from the purchaser with the aim of continuing to conduct research at this complex indefinitely. Venter noted, "We are pleased to be working with BioMed as a real estate partner as will enable us to remain financially sound for the foreseeable future, and ensure that we can continue our groundbreaking genomic science."[49]

Before considering Venter's for-profit unit, we turn and consider his nonprofit team's synthetic biology research progress in the areas of basic genomic and environmental genomic research.

Notes

1. Meredith Wadman, "Biology's ad Boy is Back," *Fortune* 149, no. 5 (March 8, 2004): 167–76, at 174, 176.
2. J. Craig Venter, *A Life Decoded: My Genome: My Life* (New York: Penguin, 2007), 331. See also Antonio Regaldo, "Next Dream for Venter," *Wall Street Journal*, June 29, 2005, A1.
3. For the concept and significance of disruptive innovation, see Joseph L. Bower and Clayton M. Christensen, "Disruptive Technologies: Catching

the Wave," *Harvard Business Review* 73, no. 1 (January–February 1995): 43–53.

4. Quoted in Thomas Kupper, "Deal Blooms for Algae Biofuel Research," *San Diego Union-Tribune*, July 15, 2009, A1. See also The Richard Dimbleby Lecture 2007: Dr. J Craig Venter—A DNA-Driven World, December 4, 2007, http://www.bb.co.uk/print/pressoffice/pressreleases/stories/2007/12_december/05/dimble (accessed August 14, 2009).

5. Quoted in Nicholas Wade, "Thrown Aside, Genome Pioneer Plots a Rebound," *New York Times*, April 30, 2002, F1.

6. Venter, *Life Decoded*, 220.

7. Kowalski Communications, Press Release, "J. Craig Venter, Ph.D., Announces Formation of Three Not-for-Profit Organizations," April 30, 2002; Venter, *Life Decoded*, 334. See also Justin Gillis, "A New Outlet for Venter's Energy," *Washington Post*, April 30, 2002, E1; Wade, "Thrown Aside"; Bruce Lieberman, "Human Genome Researcher has Plans for New Lab," *San Diego Union-Tribune*, October 7, 2002, B2.

8. J. Craig Venter Science Foundation, 2004 Internal Revenue Service Form 990, Statement of Organization's Primary Exempt Purpose—Part III; Kowalski Communications, Press Release "J. Craig Venter, Ph.D., Announces Formation."

9. Kowalski Communications, Press Release, "J. Craig Venter, Ph.D., Announces Formation."

10. Ibid. and Kowalski Communications, "IBEA Receives $3 Million Dept of Energy Grant for Synthetic Genome Development," November 21, 2002.

11. Institute for Biological Energy Alternatives (IBEA), 2002 IRS Form 990, Statement of Organization's Primary Exempt Purpose Part III—Exploration; Kowalski Communications, Press Release, "J. Craig Venter, Ph.D., Announces Formation."

12. Venter, *Life Decoded*, 334.

13. Ibid., 334–35. [Press Release] J. Craig Venter Science Foundation Joint Technology Center Inc. (Joint Technology Center), 2003 Internal Revenue Service Form 990 Statement of Program Service Accomplishments, Statement 1.

14. Joint Technology Center, 2003 IRS Form 990, Part I Revenue, Expenses, and Changes in Net Assets or Fund Balances and The Institute for Genomic Research (TIGR), IBEA, The Center for the Advancement of Genomics (TCAG), Press Release, "TIGR, IBEA, and TCAG to Create New High-Throughput Genomic Sequencing Facility," August 15, 2002.

15. Quoted in TIGR, IBEA, TCAG, Press Release, "TIGR, IBEA, and TCAG to Create."

16. Kowalski Communications, Press Release, "J. Craig Venter Announces Consolidation of Three Research Organizations Into One New Not-for-Profit Organization—The J. Craig Venter Institute," September 29, 2004; J. Craig Venter Science Foundation Joint Technology Center, 2004 IRS Form 990, Statement 6A, Articles of Merger, The Center for the Advancement of Genomic, Inc., Institute for Biological Energy Alternatives, Inc., and J. Craig Venter Science Foundation Joint Technology Center, Inc., August 30, 2004.

17. Kowalski Communications, Press Release, "The Institute for Genomic Research (TIGR), J. Craig Venter Institute, J. Craig Venter Science Foundation Consolidate into one Organization—the J. Craig Venter Institute, October 16, 2006; see also J. Craig Venter Institute, 2006 IRS Form 990, Articles of merger merging the Institute for Genomic Research, Inc. and J. Craig Venter Science Foundation, Inc. into J. Craig Venter Institute, October 1, 2006; Venter, *Life Decoded*, 356.

18. J. Craig Venter Institute, Press Release, "J. Craig Venter Institute Announces Management Team and Organization Structure," April 11, 2007.

19. Quoted in Ibid.

20. Venter Institute, 2008 and 2007 IRS Forms 990, Part I Revenue Expenses, and Changes in Net Assets or Fund Balances and Venter Institute, About the J. Craig Venter Institute, http://www.jcvi.org/cms/about/overview (accessed August 14, 2009).

21. Venter Institute, *Genomic Medicine*, http://www.jcvi.org/cms/research/groups/genomic-medicine (accessed August 14, 2009).

22. Venter, *Life Decoded*, 342.

23. Ibid., 331.

24. J. Craig Venter Institute, 2005 IRS Form 990, Schedule A, Part III—Explanation for line 2A; J. Craig Venter Institute, 2006 and 2007 IRS Form 990, Self Dealing Statement.

25. Venter Institute, *Microbial and Environmental Genomics*, http://www.jcvi.org/cms/research/groups/microbial-environmental-genomics> (accessed August 14, 2009).

26. Venter Institute, *Synthetic Biology and Bioenergy*, http://www.jcvi.org/cms/research/groups/synthetic-biology-bioenergy (accessed August 14, 2009).

27. Jim Lane, "God Loses Monopoly: Synthetic Genome Creates First Synthetic Bacterial Cell," *Biofuels Digest*, May 21, 2010, http://www.biofuelsdigest.com (accessed May 24, 2010).

28. IBEA, Press Release, "Energy Department Awards $9 Million for Energy Related Genomic Research," April 24, 2003.

29. IBEA, Press Release, "IBEA Receives $3 Million Department of Energy Grant for Synthetic Genome Development," November 21, 2002; U.S. Department of Energy, Office of Science, Genomics: GTL Research Awards, "Research to Better Understand microbial Communities and to Develop New, Biological Methods to Capture Carbon Dioxide from the Atmosphere and to Produce Hydrogen," n.d., http://genomicsgtl.energy.gov/research/ibea.shtml (accessed September 23, 2009).

30. Venter, *Life Decoded*, 350.

31. Quoted in IBEA, Press Release, "IBEA Receives $3 Million Dept. of Energy Grant."

32. IBEA, Press Release, "Energy Department Awards $9 Million," and U.S. Department of Energy, Office of Science, Press Release, "Energy Department Awards $9 Million for Energy Related Genomic Research," April 23, 2003.

33. TIGR, Press Release, "TIGR NIAIA Sign $65 Million Microbial Sequencing Contract," October 2, 2003.

34. Venter Institute, Press Release, "J. Craig Venter Institute Awarded $43 Million Five Year Contract from NIAID to Continue to Develop and Provide Sequencing, Genotyping, and Bioinformatics Expertise and Services in Infectious Diseases," May 29, 2009.
35. Ibid.
36. National Human Genome Research Institute, Press Release, "NHGRI Funds Next Generation of Large-Scale Sequencing Centers," November 7, 2003.
37. U.S. Department of Health and Human Services (HHS), National Institutes of Health (NIH), NIH News, "NIH Launches Human Microbiome Project," December 19, 2007.
38. HHS, NIH, NIH News, "NIH Expands Human Microbiome Project," June 23, 2009.
39. Venter Institute, Press Release, "Venter Institute Scientists, Along with Consortium Members of the NIH's Human Microbiome Project, Sequence 178 Microbial Reference Genomes Associated with the Human Body," May 20, 2010; HHS, NIH, NIH News, "NIH Human Microbiome Project Researchers Publish First Genomic Collection of Human Microbes, May 20, 2010; Human Microbiome Jumpstart Reference Strains Consortium, "A Catalog of Reference Genomes from the Human Microbiome," *Science* 328, no. 5981 (May 21, 2010): 994–99.
40. IBEA, Press Release, "IBEA Researchers Publish Results from Environmental Shotgun Sequencing of Sargasso Sea," March 4, 2004; Gordon and Betty Moore Foundation, Press Release, "Gordon and Betty Moore Foundation [G]ive IBEA $4.25 Million Grant for Genomic Sequencing of DNA Samples from Expedition," March 4, 2004. See also Venter, *Life Decoded*, 345.
41. Gordon and Betty Moore Foundation, Press Release, "Grants Awarded, J. Craig Venter Institute, Moore Microbial Genome Sequencing Project," n.d.; Venter Institute, Press Release, "Venter Institute to Sequence More Than 100 Key Marine Microbes in One Year," February 24, 2005.
42. Venter Institute, Press Release, "UC San Diego Partners with Venter Institute to Build Community Cyberinfrastructure for Advanced Marine Microbial Ecology Research and Analysis," January 17, 2006. Sea also Bruce Lieberman, "UCSD Scientists Join Massive Ocean Study," *San Diego Union-Tribune*, January 18, 2006, B1.
43. Quoted in Venter Institute, Press Release, "UC San Diego Partners with Venter Institute."
44. Venter Institute, Press Release, "Venter Institute Launches the J. Robert Beyster and Life Technologies 2009-2010 Research Voyage of the Sorcerer II Expedition," March 18, 2009.
45. Quoted in Ibid.
46. Venter Institute, 2007 IRS Form 990, Part III-Program Service Accomplishments and Venter Institute, 2008 IRS Form 990, Schedule 1, Part IV Business Transactions Involving Interested persons.
47. Venter Institute, Press Release, "JCVI Scientists Publish First Bacterial Genome Transplantation Changing One Species to Another," June 28, 2007; Venter Institute, Press Release, "J. Craig Venter Institute Researchers Publish Significant Advance in Genome Assembly Technology," December 4, 2008; Venter Institute, Press Release, "J. Craig Venter Institute Researchers

Clone and Engineer Bacterial Genomes in Yeast and Transplant Genomes Back into Bacterial Cells," August 20, 2009; Synthetic Genomics, "Synthetic Genomics Inc. Applauds the Venter Institute's Work in Creating the First Synthetic Bacterial Cell," May 19, 2010.

48. Venter Institute, Press Release, "J. Craig Venter Institute Sells Buildings on Rockville Campus to BioMed Realty Trust for $53 Million," May 4, 2010; BioMed Realty Trust, Inc., Press Release, "BioMed Realty Trust Acquires Five Life Science Buildings in Rockville, Maryland Totaling 218,000 Square Feet," May 4, 2010.

49. Quoted in Venter Institute, Press Release, "J. Craig Venter Institute Sells Buildings."

6

Research Progress

Continuing to build on his team's proven track record of making scientific breakthroughs, with his nonprofit organizational structure in place and funded, Venter moved on two fronts: basic genomic and environmental genomic research considered in this chapter. The practical applications of the basic genomic research advances by Venter nonprofit units helped lay the foundation for his for-profit arm, SGI, discussed in Chapter 7.

On the basic genomic research front, Venter and his team continued to pursue his ultimate goal of creating synthetic life. Having begun in 1995 with the genomic sequencing of the bacterium *Mycoplasma genitalium*, the organism with the smallest known genome capable of sustaining life, Venter sought to identify the basic genes needed for independent existence and create a new creature utilizing only these genes. His ultimate goal was the customization of life to suit the needs of the designer, whether it be biofuel, vaccine, or food production. This chapter discusses his research successes and failures on the path toward artificial life, detailing the creation of the world's first cell controlled by a synthetic genome.

Hand in hand with his work on basic genomics, Venter also dove into the oceans to discover the secrets of earth's least known habitats, in an effort "to put everything Darwin missed into context."[1] Venter cataloged millions of oceangoing microorganisms and their unique genomes in the hopes that the information may be useful not only for understanding ocean habitats, but also for producing specially designed synthetic species with metabolic pathways not seen on earth's surface. His work in environmental genomics was just as pioneering as his work in basic genomics, and is also discussed in this chapter.

Basic Genomic Research

Venter team's progress in the basic genomic research area unfolded in four steps. To date, although successful in creating a synthetic bacterial cell, Venter has not achieved the genesis of the world's first artificial life form.

The first began at the-then IBEA, led by Hamilton Smith. In November 2003, IBEA announced that it had synthesized biologically active phi-X174, a virus, more technically, a bacteriophage, that infects bacteria such as *E. coli*, but is not harmful to humans, animals, or plants.[2] The creation of synthetic phi X represented the first synthesis of artificial quasi-life. Smith's team succeeded in producing an artificial virus, based on a real one, but because a virus cannot reproduce on its own, instead relying on invasion of a foreign cell, the virus did not possess all of the ingredients that scientists consider necessary for true life. Therefore, while Smith succeeded in creating the first synthetic virus, IBEA had yet to create a life form.

To create the artificial virus, previously in the summer of 2003, some twelve months after IBEA's formation, its researchers bought commercially available short, single strands of DNA, technically, oligonucleotides, synthetically produced to match a bit of the phi X genome. After carefully purifying the individual pieces and using precise markers at both ends of each DNA piece so that they would connect to the appropriate next piece, researchers coaxed the pieces together, using polymerase cycle assembly (PCA) technology, to form a synthetic duplicate of the phi X genome. An adaptation of the PCR process (discussed in Chapter 3), PCA is a technique that produces double-stranded copies of entire individual gene sequences based on single-strand oligonucleotide, short DNA fragment templates.[3] After creating a synthetic genome of the virus, containing 5,386 base pairs scientists implanted it in a cell. The virus became active and went to work reproducing itself, just as naturally occurring virus DNA would. "Synthesis of phi X... is an important step toward our ultimate goal of synthesizing a complete cellular genome," stated Venter. He continued,

> Work in creating a synthetic chromosome/genome will at its most basic level give us a better understanding of basic cellular processes. Genome composition, regulatory circuits, signaling pathways and numerous other aspects of organism gene and protein function will be better understood through construction of a synthetic genome. Not only will this basic research lead to better understanding of these pathways and components in the particular organisms IBEA scientists are working on, but also better understanding of human biology. The ability to construct synthetic genomes may lead to

extraordinary advances in our ability to engineer microorganisms for many vital energy and environmental purposes.[4]

More specifically, the research opened the possibility of creating designer microbes to deal with excess carbon dioxide (technically, carbon sequestration) or meet future fuel needs, likely, it was thought at that time, in the form of hydrogen.

Leveraging their synthetic genome expertise, Venter's team next focused on inserting a full bacterial chromosome into a recipient cell and transforming the cell's operating system to work off the new chromosome. But this was not a simple task, and bringing this prospect to fruition took several years.

The second step toward the creation of a synthetic genome culminated in June 2007, when Venter's team successfully transplanted naturally occurring chromosomes of one species of bacterium *Mycoplasma mycoides* (*M. mycoides*) into another species of the bacterium *Mycoplasma capricolum* (*M. capricolum*), by replacing one organism's genome with another's genome.[5] In order to easily identify cells which had been transformed by the *M. mycoides* genome, researchers added a certain type of gene to a *M. mycoides* large colony chromosome which would turn the transformed cells blue. The team then purified this large colony chromosomal DNA from the *M. mycoides* so that it was free from structural proteins. Researchers took the "naked" *M. mycoides* DNA and placed it into a second bacterial cell, *M. capricolum*. The transplanted *M. mycoides* DNA changed the species from the original, *M. capricolum*, to the species defined by the transplanted DNA, *M. mycoides*. In other words, the transferred genome some one million genetic letters long, took over the living host cell, with its chromosomes spurring the machinery of life into action after landing in an unfamiliar cell. The host cell switched over to producing proteins specified by the inserted DNA and thereby changed its species mid-life.

This breakthrough demonstrated that the genome of one bacterial species could be injected into the cell of another, yielding a living cell that produced whatever its new genome told it to do. By exchanging genomes between organisms, changes could be forced by deliberate human design and selection, not merely by evolution and random selection. Humans, not nature, became the master of life.

In terms of commercial applications, businesses could use this technology to inject artificial genomes into cells to produce new energy sources, pharmaceuticals, or chemicals. In commenting on the importance of the successful completion of this research, Venter stated, "...It is one of the key proof of principles in synthetic genomics that will allow us to realize the ultimate goal of creating a synthetic organism.[6] The development of a synthetic genome soon followed.

The third step occurred in 2008, when Venter's team successfully manufactured a synthetic genome in the laboratory.[7] The first artificial genome, designed on a computer, was made chemically using very large DNA molecules, stitched together to form the entire genome with the use of any natural DNA.

In arriving at this milestone, Venter revisited his 1995 determination, as discussed in Chapter 4, of the precise order of about 580,000 DNA base pairs of *Mycoplasma genitalium* (*M. genitalium*), a bacterium that can infect the human genital tract. The complete DNA of this bacterium is more complex than the DNA of a virus, such as phi-X, previously synthesized in 2003. Venter used his work from 1995 as a template for the sequence of A, C, G, and T in his first synthetic chromosome, converting the template into a digital format which could be read and edited by computer software.

After converting *M. genitalium's* code into 1's and 0's, researchers built from scratch an entire microbial chromosome, a loop of synthetic DNA, carrying all the instructions needed for a cell to live and reproduce. Beginning with bottles of the off-the-shelf chemical ingredients (sugars, nitrogen-based compounds, and phosphates), scientists assembled more than one half million of the A's, C's, G's, and T's of *M. genitalium's* DNA in the correct order, first synthesizing DNA cassettes, or smaller pieces of the whole chromosome, and then assembling the individual cassettes into the entire chromosome, yielding a human-made *M. genitalium* genome. In short, researchers constructed a synthetic genome in the laboratory based on *M. genitalium* and its 482 genes, the smallest known genome in an independent living organism.

To obtain the cassettes, Venter's team worked with three DNA synthesis companies, Blue Heron Biotechnology, DNA 2.0, and GENEART. These companies, the foundries of the biotechnology era, make custom genes-to-order thousands of bases long by splicing together shorter strings of base pairs. The Venter researchers ordered about one hundred of such sequences, each four thousand to seven thousand base pairs in length, from these companies. They then joined these sequences, the cassettes, into ever-bigger pieces in a five-stage assembly process.[8] Eventually, four big pieces were put into yeast cells, which hooked them together using a natural gene repair mechanism.

In the first stage of the assembly process, researchers joined sets of four cassettes to create twenty-five subassemblies, each with about twenty-four thousand base pairs. These fragments were then cloned in the bacterium, *E. coli*, to produce sufficient DNA for the next steps and for DNA sequence validation. Stage two involved combining three twenty-four thousand fragments to form eight assembled blocks, each with about seventy-two thousand base pairs. These fragments were again cloned into *E. coli* for DNA production and sequencing. Stage three again involved combining these fragments to produce larger fragments of about 144,000 base pairs. Because the *E. coli* had difficulty assembling larger DNA segments, in the fourth stage, the team, used yeast, rather than *E. coli*, and a technique called homologous recombination, a process that cells naturally use to repair chromosome damage. The yeast rapidly built the entire synthetic bacterial chromosome from the subassemblies, thereby assembling the final genome of more than 580,000 base pairs. Stage five sequenced the final chromosome to validate its complete chemical structure. The project demonstrated the feasibility of building large genomes, but the multistage process was highly complex and tedious.

The final product, with 582,970 base pairs of DNA, the chemical units of the genetic code, was nearly an exact replica of *M. genitalium's* genome, with a few intentional differences inserted to differentiate the synthetic code from a naturally occurring chromosome. Specifically, Venter's team disrupted one gene so that any new microbe created using this genome could not infect other cells and added extra DNA as watermarks, specially designed DNA segments that use the "alphabet" of genes and proteins to spell out words, sentences, and numbers. Venter's effort produced basically a natural *M. genitalium*, not a new species.

For the first time, scientists synthetically replicated the entire genome of a bacterium, something that was considered alive. However, a DNA molecule alone was not itself alive. The molecule needed a cell in order to interpret its commands, and so Venter's team attempted to transplant their synthetic *M. genitalium* chromosome out of yeast and into a bacterial cell. However, all attempts to do so met with failure, and Venter's team went back to the drawing board.[9] The team would need to wait about two more years until achieving that transplantation breakthrough.

During the fourth step, in 2008 and 2009, undaunted by their past failures, Venter researchers focused on accelerating the synthetic DNA assembly process. The five-stage assembly process, described earlier, originally used to synthesize the synthetic *M. genitalium* genome was slow and laborious, and Venter's team sought a better assembly mechanism while simultaneously working to solve the transplantation problems they encountered with synthetic *M. genitalium*. In December 2008, the team published a paper describing the assembly of an *M. genitalium* genome in only one stage using twenty-five DNA fragments, rather than five separate stages.[10] In this study, scientists used DNA fragments, or cassettes, synthesized and cloned in *E. coli*, that ranged in size from seventeen thousand to thirty-five thousand base pairs. These segments were inserted into yeast cells in only one step, skipping over some of the previous stages using *E. coli*. As it had done before, the yeast assembled the cassettes through homologous recombination, creating the synthetic genome. Subsequent experiments confirmed that all twenty-five pieces of synthetic DNA were correctly assembled.

The simplification of DNA assembly technology lowered the cost of chromosome construction, potentially opening the door for the synthesis of longer DNA strands. In the future, various combinations of DNA molecules and genetic pathways could be manufactured in yeast, thereby turning yeast into a genetic factory.

Research was bogged down, however, largely because *M. genitalium* grew so slowly, resulting in a single experiment taking weeks to complete. Venter's team turned to the faster growing bacterium, *M. mycoides*, and began to build a synthetic copy of its chromosome from scratch, scrapping the work already completed with *M. genitalium*. As an added bonus, because Venter's team already had experience with transplanting the *M. mycoides* genome, they needed to spend far less time learning how to transfer the synthetic genome into its new cellular host.

In August 2009, Venter researchers published results describing new methods of cloning the complete genome of *M. mycoides* in a yeast cell, without the need for first synthesizing and cloning cassettes in *E. coli*, marking the first time a bacterial genome had grown successfully in yeast.[11] Building on research published in 2002, the team cloned the bacterial genome in yeast, choosing to do so because of yeast's ability to handle larger DNA molecules than *E. coli*, modifying the genome as if it were a yeast chromosome. They then transplanted the genome back into a recipient bacterial cell of a related species, *M. capricolum*, to create a new bacterial strain, a new type of *M. mycoides* cell. The ability to modify bacterial genomes in yeast extended yeast genetic tools to the bacterial world. In noting this work had important implications for understanding biological fundamentals and for creating and booting up a synthetic genome, Hamilton Smith stated, "This is possibly one of the most important new findings in the field of synthetic genomics."[12]

The fifth step toward synthetic life took place in May 2010, when the Venter–Smith team announced the momentous culmination of their quest to create the first synthetic bacterial cell. Researchers outlined the steps to synthesize a 1.08 million base pair *M. mycoides* genome, a single cell organism, constructed from bottles of off-the-shelf chemicals that make up DNA, turning computer code into a living organism.[13] The 1.08 million base pair synthetic genome was the largest chemically defined structure ever synthesized in the laboratory. The genome was booted up in a cell to create the first cell controlled by an artificial genome. In contrast to the Venter team's 2009 creation of the first bacterial genome assembled entirely by yeast, this *M. mycoides* bacterial genome, while also assembled in yeast, was not synthesized by cloning natural *M. mycoides* DNA, but rather was synthesized using only the 1's and 0's of computer code.

Creating a synthetic bacterial genome and using it to control a cell involved three basic steps: assembly; insertion; and self-replication. After starting with the digital computer code of *M. mycoides*, the team designed and ordered 1,078 specific pieces (cassettes) of DNA 1,080 base pair units in length from Blue Heron Biotechnology. Researchers ordered these short segments because DNA synthesizing technology continued to lag behind that for DNA sequencing. According to Clyde Hutchison, Ph.D., head of the SGI's scientific advisory board, "To me the most remarkable thing about our synthetic cell is that its genome was designed in the computer and brought to life through chemical synthesis, without using any pieces of natural DNA."[14]

Researchers then used a previously developed technique for assembling the shorter lengths of DNA into a complete genome, a loop one million units in length, thereby turning the digital code into a series of DNA genetic sequences. To accomplish stitching the pieces together, the team employed a three-stage process using a yeast assembly system. The first stage involved taking 10 cassettes of DNA at a time to build one hundred ten 10,000 base pairs segments. In the second

91

stage, the 10,000 base sequences were taken ten at a time to produce eleven 100,000 base pairs segments. Finally, all 100,000 base sequences were assembled into the complete synthetic genome in yeast cells and grown as a yeast artificial chromosome. The completed loop of DNA was designed to replicate the *M. mycoides* genetic sequence.

The team isolated the synthetic genome from the yeast cells and, as they had done with a naturally occurring bacterial chromosome in 2007, inserted it into recipient cells of a different, but closely related, bacterium species, *M. capricolum* that had the genes for its restriction enzyme removed. Restriction enzymes are DNA proteins found in most bacteria that cut DNA molecules at specific points, and are thought to have developed as a defense against foreign DNA, such as viruses. Restriction enzymes will not cut the cell's own DNA, however, because that DNA is protected by another bacterial enzyme. Originally, Venter's team was unable to transplant the synthetic genome from yeast into *M. capricolum* because yeast, which is a fungus instead of a bacterium, does not use the protection enzyme for DNA inside itself. Venter's team solved this problem by inactivating *M. capricolum's* restriction enzyme so that it would not attack the synthetic genome.[15]

After the synthetic DNA was inserted, it was transcribed into mRNA by *M. capricolum's* own cellular machinery that, in turn, was translated into new proteins at *M. capricolum's* own ribosomes. Put simply, the *M. mycoides* genome used *M. capricolum* as the architect of its own demise. Although the host *M. capricolum* genome was destroyed by the *M. mycoides* restriction enzymes or was lost during cell replication, the host cells were equipped with proteins created by the original *M. capricolum* genome that could make sense of the new genome the cells received. Already possessing the ability to read their new instructions, the cells began to act according to the command of the newly inserted synthetic *M. mycoides* genome, losing their previous identity.

After two days, viable *M. mycoides* cells, containing only DNA, took over the *M. capricolum* cells and substituted for the host cell's own DNA. The artificial genome then replicated itself to form a bacterial colony, making proteins characteristic of *M. mycoides*. Venter stated, "We think these are the first synthetic cells that are self-replicating and whose genetic heritage started in the computer."[16]

As they had done with *M. genitalium* in 2008, researchers again designed and inserted watermarks into the *M. mycoides* synthetic genome. An essential means to prove a genome is of human creation and to identify a laboratory of origin; the Venter team chose to mark the synthetic genome with their names, a web address, and three quotations.[17] Researchers also modified the genome so as to remove any potential danger posed by accidental release of microbes. The synthesized genome was based on the *M. mycoides* bacterium that infects goats, and the two were nearly identical. But before using that bacterium as a template, researchers digitally excised those genes believed to be pathogenic, about fourteen of 850 total genes, so that if the genome escaped, it would not harm goats.

Looking toward the future, scientists hope that the knowledge gained from constructing the first self-replicating synthetic cell, together with decreasing costs for DNA synthesis, will facilitate the wide scale use of synthetic biotechnology. Although the bacterium, *M. mycoides*, is not suitable for biofuels production, Venter indicated that he would move to different, more complex species, notably, algae, having a genome in the nucleus of its cells. He said that he would attempt to build "an entire algae genome so we can vary the 50 to 60 different parameters for algae growth to make super-productive organisms."[18]

The Venter team will also work on its ultimate objective of synthesizing a minimal artificial cell containing only the genes needed to sustain life in its simplest form. Once they achieve this goal, they would proceed to develop increasingly novel designer genomes.

The ability to synthesize large strands of DNA opened the door for the creation of customized life. By combining this new technology with his 1995 discovery of the minimal set of genes needed for survival and reproduction, about three hundred in total, Venter envisions a microbe carrying a minimalist genome that could be used as a scaffold onto which new genes could be added to create organisms engineered to perform specific functions. In the near future, a new artificial creature may enter the world, perhaps shedding light on the last common ancestor of earthly life. However, it is unclear which of the 480 or so genes in the *M. genitalium* genome, the genome that allowed Venter to determine the minimum set needed for life, can be eliminated without killing the bacterium, because the genes' interrelations and total functions are not yet fully understood.

The final step, the creation of a new life form, has not been accomplished to date. Venter and his team still cannot design completely new organisms from scratch, nor can they create any synthetic organism without the use of something already alive. Researchers from around the world have likened Venter's forced transformation of a cell to taking a Mac operating system, installing it on a PC, and the PC becoming a Mac.[19] Venter himself acknowledged that his team's success depended on their use of a full-grown adult bacterial cell, remarking that he and his team were "taking advantage of three and a half billion years of evolution."[20]

By building on Venter's successes, however, the design process of the future could turn inanimate chemicals into a customized living organism. This process would involve the creation of a living bacteria cell, a primitive microbe, based entirely on a synthetically made genome represented as 1's and 0's in a computer. Researchers would insert this synthetic genome into some type of living "shell," have it boot up, and take control of the host's functions, thereby creating a new form of life from scratch, not mimicking an existing creature.

In addition to making a great technological leap forward, the new organism might also blur or erase the distinction between living and nonliving matter. Ultimately, scientists expect to use the ability to create new cells with new functions to build bigger, more complex species crafted to pump out fuels, chemicals, pharmaceuticals, and even food products more cheaply and efficiently than is presently possible.

Environmental Genomic Research

Venter, a man of diverse interests, actively pursued the area of environmental genomic research, exploring untapped microorganisms from diverse marine environments. His team discovered novel genes and pathways residing in the genomes of these microorganisms that were previously unknown to humans. Genomic-driven molecular tools, it is expected, would provide new ways to evaluate microbial metabolism, ecology, evolution, biochemistry, physiology, and biodiversity.

Beginning in 2003, Venter's team went prospecting for new organisms in the deep mid-ocean, too far down for solar energy, too high up for geothermal energy. Awash in novel DNA, the initial expedition uncovered new genes, species, and protein families.

As a pilot study, Venter set sail from Bermuda in February and May 2003 to build a genomic catalog of the ocean's microbes. Combining work with pleasure aboard his yacht, the Sorcerer II, Venter stopped at the Sargasso Sea, near Bermuda, one of the most studied regions of the earth's oceans. Long presumed to be relatively devoid of microbial life, Venter captured bacteria and viruses, microorganisms not easily grown in lab dishes, on progressively smaller filters and then shipped the frozen, filtered samples back to IBEA laboratories. There, scientists extracted the organisms' genomic DNA and made DNA libraries. They analyzed these organisms using the long-perfected whole genome shotgun sequencing method; the same technique used to sequence the human genome, and assembled the sequence results virtually with precise mathematical algorithms, among other computational tools.

Revealing a diversity of ocean life previously unseen, Venter's research in the first half of 2003 uncovered more than 1.2 million new genes and at least 1,800 new microbial species.[21] The team also found nearly six hundred new genes (so-called photoreceptor genes) for light sensitive proteins, suggesting more bacteria may convert light into other types of energy. According to Venter, "It is estimated that over 99% of species remain to be discovered. Our work in the Sargasso Sea, an area thought to have low diversity of species, has shown that there is much that we do not yet understand about the ocean and its inhabitants."[22]

Subsequently, in a two-year circumnavigation of the globe, beginning in August 2003, the Sorcerer II collected samples of sea water every two hundred nautical miles. Venter's team discovered more than 6 million new genes during the voyage, doubling the total in the globe's genetic databases, and 1,700 unique protein families, particularly new protein kinases, some of the most important cellular proteins.[23] Protein kinases are enzymes, a type of protein that regulates many basic cellular functions in all species, including humans, by attaching phosphate chemical groups to other proteins and small molecules. For instance, some of the new proteins protect microbes from ultraviolet ray damage while others help repair the damage caused by ultraviolet light.

To handle the enormous volume of data generated by the voyage between Halifax, Nova Scotia and French Polynesia, the team developed new computational methods to assemble and analyze the microbial DNA. Fragment recruitment, a comparative genomic method, allowed scientists to examine genome structure, microbial evolution, and species diversity on many levels. New assembly techniques enabled researchers to bring together large DNA segments from abundant, but previously hard to analyze, genomes of organisms. These new tools, among others, enhanced our knowledge of the biological processes of microbial communities, thereby helping unlock the mysteries of unseen life.

With the rate of discovery of new genes and proteins as great at the end of the voyage as at the start, humanity is probably not even close to cataloging the totality of global biodiversity. "Instead of being at the end of discovery, it means we're in the earliest stages," noted Venter.[24]

The diverse supply of oceanic microbial DNA may comprise a rich lode for both scientists and businesses. Having analyzed the DNA of many seawater microbes, Venter has accumulated a library of some forty million genes, mostly from algae.[25] He hopes to use these genes as a resource to make algae produce fuels and chemicals, among other products. Other researchers at pharmaceutical firms will hunt for new compounds in sea creatures that can be used in drug applications. The data may also be used to compare the DNA of ocean microorganisms to bacteria and viruses that cause human diseases in order to develop more effective preventative medicine, such as vaccines. The findings could also help pave the way to alternative energy sources. By adding genes from sea organisms, microbes created in the laboratory may be engineered to release hydrogen gas, among other energy possibilities.

On the for-profit side, Venter's goal (as well as that of other firms), focuses on using his team's research to design novel microbes with handcrafted genomes that endow them with the ability to produce useful products, such as fuels and chemicals. The next chapter considers his efforts to produce a genetic platform able to direct the basic functions of life and then attach custom-designed DNA modules to compel cells to make biofuels.

Notes

1. Quoted in Tim Adams, "The Observer Profile: Craig Venter: The First Lord of the Laboratory," *Observer* (England), May 23, 2010, 26.
2. Hamilton O. Smith et al., "Generating a Synthetic Genome by Whole Genome Assembly: øX174 Bacteriophage from Synthetic Oligonucleotides," *Proceedings of the National Academy of Sciences* 100, no. 26 (December 23, 2003): 15440–45. See also Institute for Biological Energy Alternatives (IBEA), Press Release, "IBEA Researchers Make Significant Advance in Methodology Toward Goal of a Synthetic Genome," November 13, 2003; J. Craig Venter, *A Life Decoded: My Genome: My Life* (New York: Viking Penguin, 2007), 350–53; Elizabeth Pennisi, "Venter Cooks up a Synthetic Genome in Record Time," *Science* 302, no. 5649 (November 21, 2003): 1307; Elizabeth Weise, "Scientists Create a Virus that Reproduces," *USA*

Today, November 14, 2003, 5A; Rick Weiss, "Researchers Create a Virus in Record Time," *Washington Post*, November 14, 2003, A10; John J. Fialka, "Science Advances on Bacteria Made to Fight Pollution," *Wall Street Journal*, November 14, 2001, A4.

3. For a technical discussion of the PCA process, see William P. C. Stemmer et al., "Single-Step Assembly of a Gene and Entire Plasmid from Large Numbers of Oligodeoxyribonucleotides," *Gene* 164, no. 1 (October 16, 1995): 49–53.

4. Quoted in IBEA, Press Release, "IBEA Researchers."

5. Carole Lartigue et al., "Genome Transplantation in Bacteria: Changing One Species to Another," *Science* 317, no. 5838 (August 3, 2007): 632–38. See also J. Craig Venter Institute (Venter Institute), Press Release, "JCVI Scientists Publish First Bacterial Genome Transplantation Changing One Species to Another," June 28, 2007; Nicholas Wade, "Pursuing Synthetic Life, Scientists Transplant Genome of Bacteria," *New York Times*, June 29, 2007, A1; Rich Weiss, "Scientists Report DNA Transplant," *Washington Post*, June 29, 2007, A3; Gautam Naik, "J. Craig Venter's Next Big Goal," *Wall Street Journal*, June 29, 2007, B1; Elizabeth Pennisi, "Replacement Genome Gives Microbe New Identity," *Science* 316, no. 5833 (June 29, 2007): 1827.

6. Quoted in Venter Institute, Press Release, "JCVI Scientists."

7. Daniel G. Gibson et al., "Complete Chemical Synthesis, Assembly, and Cloning of a *Mycoplasma gentialium* Genome," *Science* 319, no. 5867 (February 29, 2008): 1215–20. See also Venter Institute, Press Release, "Venter Institute Scientists Create First Synthetic Bacterial Genome," January 24, 2008; Drew Endy, "Reconstruction of the Genomes," *Science* 319, no. 5867 (February 29, 2008): 1196–97; Rick Weiss, "Md. Scientist Build Bacterial Chromosome," *Washington Post*, January 25, 2008, A4; Andrew Pollack, "Researchers Announce a Step toward Synthetic Life," *New York Times*, January 25, 2008, A15; Gautam Naik, "Scientists Advance in Effort to Create Synthetic Organism," *Wall Street Journal*, January 25, 2008, B3; "Nearly There," *Economist* 386, no. 8564 (January 26, 2008): 76–77; Justin Gillis, "Scientists Planning to Make New Form of Life," *Washington Post*, November 21, 2002, A1.

8. For a technical discussion of the assembly process, see Yogesh S. Sanghvi, "A Roadmap to the Assembly of Synthetic DNA from Raw Materials," http://dspace.mit.edu/handle/1721.1/39657 (accessed May 25, 2010).

9. Venter Institute, Press Release, "J. Craig Venter Institute Researchers Clone and Engineer Bacterial Genomes in Yeast and Transplant Genomes Back into Bacterial Cells," August 20, 2009.

10. Daniel G. Gibson et al., "One-Step Assembly in Yeast of 25 Overlapping DNA Fragments to Form a Complete Synthetic *Mycoplasma genitalium* Genome," *Proceedings of the National Academy of Sciences* 105, no. 51 (December 23, 2008): 20404–9. See also Venter Institute, Press Release, "J. Craig Venter Institute Researchers Publish Significant Advance in Genome Assembly Technique," December 4, 2008.

11. Carole Lartigue et al., "Creating Bacterial Strains from Genomes That Have Been Cloned and Engineered in Yeast," *Science* 325, no. 5948 (September

25, 2009): 1693–96. See also Venter Institute, Press Release, "J. Craig Venter Institute Researchers Clone and Engineer."

12. Quoted in Venter Institute, "J. Craig Venter Institute Clone and Engineer."

13. Daniel G. Gibson et al., "Creation of a Bacterial Cell Controlled by a Chemically Synthesized Genome," *Science*, 329, no. 5987 (July 2, 2010): 52–6; Venter Institute, Press Release, "First Self-Replicating, Synthetic Bacterial Cell Constructed by J. Craig Venter Institute Researchers," May 20, 2010; Venter Institute, First Self-Replicating Synthetic Bacterial Cell, http://www.jcvi.org/cms/press/press-releases/full-text/article/first-self-replicating-synthetic-bacterial-cell-constructed-by-j-craig-venter-institue-researchers/ (accessed May 21, 2010); J. Craig Venter and Daniel Gibson, "How We Created the First Synthetic Cell," *Wall Street Journal*, May 26, 2010, A17. See also Nicholas Wade, "Synthetic Bacterial Genome Takes Over a Cell, Researchers Report," *New York Times*, May 21, 2010, A17; Robert Lee Hotz, "Scientists Create Synthetic Organism," *Wall Street Journal*, May 21, 2010, A1; David Brown, "Creating a Cell from Scratch," *Washington Post*, May 21, 2010, A3; Elizabeth Pennisi, "Synthetic Genome Brings New Life to Bacterium," *Science* 328, no. 5981 (May 21, 2010): 958–59; Gary Robbins, "First Synthetic Cell Created by Scientists," *San Diego Union-Tribune*, May 21, 2010, A1; Gary Robbins, "Venter Shares his Vision for Applying Technology," *San Diego Union-Tribune*, May 21, 2010, A1; Katherine Bourzac, "How to Remake Life," *Technology Review* 113, no. 5 (September 1, 2010): 108–10.

14. Quoted in Venter Institute, Press Release, "First Self-Replicating."

15. For the assembly and transplantation, see Gibson, "Creation of a Bacterial Cell Controlled," 52–55.

16. Quoted in Brown, "Creating a Cell," A3.

17. Venter Institute, Press Release, "First Self-Replicating" and Natalie Angier, "Peering Over the Fortress that is the Mighty Cell," *New York Times*, June 1, 2010, D1.

18. Quoted in Wade, "Synthetic Bacterial Genome," A17.

19. Deborah Smith, "Cell Door Opens on Fantastic Future," *Sydney Morning Herald* (Australia), May 27, 2010, 18.

20. Quoted in Angier, "Peering Over the Fortress," D1.

21. J. Craig Venter et al., "Environmental Genome Shotgun Sequencing of the Sargasso Sea," *Science* 304, no. 5667 (April 2, 2004): 66–74. See also IBEA, Press Release, "IBEA Researchers Publish Results from Environmental Shotgun Sequencing of Sargasso Sea in Science," March 4, 2004. See also Venter, *Life Decoded*, 343–45; Andrew Pollack, "Groundbreaking Gene Scientist is Taking His Craft to the Oceans," *New York Times*, March 4, 2004, A19; Rick Weiss, "Md. Team Finds Species of Sea Microbes," March 5, 2004, A7; Andrew Pollack, "A New Kind of Genomics, with an Eye on Ecosystems," *New York Times*, October 21, 2003, F1.

22. Quoted in IBEA, Press Release, "IBEA Researches Publish."

23. Shibu Yooseph, et al., "The *Sorcerer II* Global Ocean Sampling Expedition: Expanding the Universe of Protein Families," *PLoS [Pacific Library of*

Science] Biology 5, no. 3 (March 2007): e16; Douglas B. Rusch et al., "The *Sorcerer II* Global Ocean Sampling Expedition: Northwest Atlantic through Eastern Tropical Pacific," *PLoS Biology* 5, no. 3 (March 2007): e77; Natarajan Kannan et al., "Structural and Functional Diversity of the Microbial Kinome," *PLoS Biology* 5, no. 3 (March 2007): e17; Venter Institute, Press Release, "More than Six Million New Genes, Thousands of New Protein Families, and Incredible Degree of Microbial Diversity Discovered from First Phase of Sorcerer II Global Ocean Sampling Expedition," March 13, 2007; Venter, *Life Decoded*, 345–48; Venter Institute, Press Release, "Venter Institute Launches the J. Robert Beyster and Life technologies 2009–2010 Research Voyage of the Sorcerer II Expedition," March 18, 2009. See also Gautam Naik, "Seafaring Scientist Sees Rich Promise in Tiny Organisms," *Wall Street Journal*, March 13, 2007, B1; Rick Weiss, "Seas Yield Surprising Catch of Unknown Genes," *Washington Post*, March 14, 2007, A1.

24. Quoted in Weiss, "Seas Yield," A1.
25. Wade, "Synthetic Bacterial Genome," A17.

7

Commercialization Efforts: Synthetic Genomics, Inc.

Building on his team's research prowess, Venter founded SGI in February 2005, a for-profit firm. In creating the new firm, he noted, "Rapid advances in high throughput DNA sequencing and synthesis, as well as high performance computing and bioinformatics, now enable us to synthesize novel photosynthetic and metabolic pathways. Using diverse sets of genes, including those from over 300 fully-sequenced genomes, will allow our new company to develop synthetic organisms for specific industrial applications." He continued, "We are in an era of rapid advances in science and are beginning the transition from being able to not only read genetic code, but are now moving to the early stages of being able to write code."[1]

After discussing SGI's cofounders and its top executives, this chapter considers the firm's mission and business plan, its funding, and its three main project areas. SGI currently works in three areas: microbial-enhanced coal recovery in collaboration with BP PLC (BP); agricultural biofuels in partnership with the Asiatic Centre for Genome Technology (ACGT); and biofuels and biochemicals production in an alliance with ExxonMobil Research and Engineering Company (EMRE). The firm partners with ExxonMobil and BP on projects that seek to use synthetic biology to make cheap, cleaner fuels from algae and coal-bed methane. For reasons ranging from a fear of running out of crude oil, concerns about global warming, to placating environmentalists, a demand exists to find new, renewable fuel sources, essentially identical to existing fossil-based fuels. SGI's algae-based fuel technologies, while still in the development stage, embody a promising solution to these concerns, and SGI's agreement with a unit of ExxonMobil represented a huge vote of confidence in Venter, SGI, and the entire algae-to-fuel industry.

SGI's scientific strength lies in the pioneering research of Venter and Smith as well as the team they assembled at the firm and the Venter Institute. In addition to its own applied research efforts and internally developed intellectual property, SGI funds research at the Venter Institute and has exclusive access to new inventions and discoveries in synthetic genomics research developed by the institute under the Sponsored Research Agreement between both organizations.[2] The firm leverages its intellectual property and internal knowledge in its partnerships with BP, ACGT, and ExxonMobil to develop new genomic advances and disruptive technologies.

Synthetic Genomics: Its Cofounders and Its Top Executives

Venter cofounded SGI with Hamilton O. Smith, Ph.D., Juan Enriquez, and David Kiernan, MD.[3] As discussed in Chapter 4, Dr. Smith received the 1978 Nobel Prize for his work on the discovery of restriction enzymes, an essential tool used in the fields of rDNA technology and genomic sequencing. A renowned expert in DNA library construction and DNA manipulation techniques, Smith left Johns Hopkins University where he was a professor of molecular biology and genetics and joined the scientific team at TIGR, a Venter nonprofit firm. Smith also served as the former senior director of DNA Resources at Celera Genomics, as discussed in Chapter 4. At SGI he serves as co-chief scientific officer, together with Venter.

Mr. Enriquez, a writer, served as the founding director of the Harvard Business School's Life Sciences Project and the founder of Biotechonomy LLC, a life sciences research and investment firm. He is also the managing director of Excel Venture Management, a life sciences and healthcare venture capital firm. Mr. Enriquez cofounded SGI and currently serves on its board of directors.

Dr. Kiernan, a physician and attorney, is a senior litigation partner at Williams & Connolly, a Washington, DC, law firm. Dr. Kiernan also cofounded SGI and currently serves on the board of directors of the Venter Institute.

After founding SGI, Venter moved quickly to bring an experienced top manager to the firm. In February 2006, SGI named Aristides A.N. Patrinos, Ph.D., as its president, a significant coup.[4] Dr. Patrinos joined the firm from the DoE's Office of Science, where he served as the associate director of the department's Office of Biological and Environmental Research. There he directed various research activities, including the DoE's human and microbial genome programs and its efforts to solve energy and environmental problems using microbes.

Dr. Patrinos was well known for his role in developing the U.S. Government-funded Human Genome Project and helping to guide the historic sequencing of the human genome to completion. As part of his leadership in the Human Genome Project, Dr. Patrinos helped create the DoE's Joint Genome Institute in 1997 to unite the department's expertise and resources in genome mapping, DNA sequencing, technology development, and information sciences as pioneered at four DoE genome centers. He also developed and launched the DoE's Genomes to Life Program, a research program dedicated to developing technologies to understand and use the diverse capabilities of microbes for innovative solutions to the challenges faced by the DoE's energy and environmental mission, as noted in Chapter 5. The Genomes to Life Program funded, in part, research projects at Venter's IBEA.

Venter also recruited a distinguished scientific advisory board for SGI. Board members include, Clyde Hutchison III, Ph.D., a molecular biology pioneer and distinguished investigator of the synthetic biology team at the Venter Institute. Hutchinson serves as the chair of SGI's board of scientific advisors.

Synthetic Genomics' Mission and Business Plan

SGI seeks to develop and commercialize genomic-driven technologies to meet the demand for critical resources, including energy and chemical products.[5] Believing a need exists for cleaner, greener energy and chemical sources, SGI centers its efforts on solutions primarily in these two areas as well as secondarily in pharmaceuticals and food oils.

To accomplish its goals, SGI uses genes as design components to develop custom modular units, in the form of DNA cassettes, which encode microbial metabolic pathways for selected commercial applications. Put simply, SGI's technology uses DNA sequences containing a number of genes that work together to convert one chemical, such as carbon dioxide, into another, such as petroleum.

At SGI, synthetic genomics involves the design and assembly of genes, gene pathways, and whole chromosomes from the chemical components of DNA. The genome of a cell serves as the cell's operating system with the cell's cytoplasm as the hardware. As discussed in Chapter 3, the cytoplasm contains ribosomes, among other components, needed for the expression of the genetic information contained in the genome. SGI seeks to modify the cell's operating system, design new genomes, and code for new types of cells with desired properties focused on production of bioenergy, substitutes for petrochemicals, food oils, and new pharmaceuticals. In other words, SGI focuses on manufacturing organisms that perform specific functions and that can be inserted into cells, modifying the cellular operating system, helping produce biofuels, among other products. In the future, it is believed that SGI's science and technology can be extended to a variety of other solutions, from human health, food, and water production to the environment.

To achieve its goals, SGI focuses on synthesizing and programming DNA by developing and using advances in synthetic genomics. It applies these advances to design and develop microbes for industrial processes and environmental applications. It harnesses the natural diversity of microbes' genomes to develop optimally useful metabolic pathways and make them more efficient on a commercial scale. SGI employs both novel and public domain computational tools to identify and annotate new genes, assemble genomic sequences, and compare closely related genomes. It also focuses on developing microbial cultivation technologies and various approaches that reveal novel microorganisms, new genes, and fermentation platforms for metabolic engineering of microbes for commercial applications.

SGI built its business model on its unique scientific knowledge as well as its proprietary tools and techniques that are applied toward product development in several key industries.[6] Its diverse solution portfolios represent both near-term, attainable commercial opportunities and longer-term market-changing, disruptive applications. According to Venter, "I believe the best examples of disruptive technologies that could change our future are in the new fields of synthetic biology, synthetic genomics, and metabolic engineering. These fields can change the way we think about life by showing that we can use living systems to increase our chances of survival as a species. Simply put: this area of research will enable us to create new fuels to replace oil and coal."[7]

Collaborations with major firms serve to validate its technology and commercial platforms. These partnerships are discussed later in this chapter.

104

Funding of Synthetic Genomics

Unlike his two previous collaborations, first with Human Genomic Sciences and then with Perkin-Elmer, in 2005 Venter did not seek a corporate sponsor-funder for SGI. Instead, Venter turned to private investors and venture capital firms.

Initially, SGI was funded with $30 million Venter raised, including a $15 million investment from Alfonso Romo Garza, a Mexican billionaire, chairman and CEO of Mexico's Pulsar, Inc., and the controller of a chunk of the world's commercial vegetable seeds.[8] As Romo stated, "Of course, [Venter's] antagonistic. He's controversial. But I love controversial people because those are the people who change the world."[9] Venter also obtained another $15 million for SGI from a small group of other wealthy individuals.

Venter then turned to venture capital firms, three project collaborators, and a biotechnology tools company.[10] In 2006, SGI raised $30 million in a Series A financing from Draper Fisher Juvetson, Meteor Group, Biotechonomy LLC (Juan Enriquez, a cofounder of SGI, is the CEO of Biotechonomy), and Plenus, SA de CV. In 2007, it closed its Series B round of financing with two project collaborators, BP and the ACGT Sdn Bhd. For an investment of some $10 million, the latter entity received 4.5 percent stake in SGI.[11] Then, in 2010, Life Technologies, a provider of innovative life science solutions, made on equity investment in SGI.[12]

Synthetic Genomics' Project Areas

SGI, now headquartered in La Jolla, California, is currently working in three project areas: microbial-enhanced coal recovery in collaboration with BP; agricultural biofuels in partnership with the ACGT; and the next generation of biofuels and biochemicals in an alliance with EMRE. This section considers each of these three project areas.

BP Collaboration

In June 2007, SGI and BP formed a multiyear research and development collaboration and commercialization joint venture to develop and bring to market microbial-enhanced solutions to increase the conversion and recovery of subsurface hydrocarbons, including coal, petroleum, and natural gas.[13] Conversion of subsurface hydrocarbons involves changing one type of hydrocarbon to another, from a solid to a gas, for instance, to make recovery of the fuel easier or more efficient. A biological conversion process for subsurface hydrocarbons, one using microbes' natural metabolic pathways, hopefully will lead to cleaner energy production and improved recovery rates.

BP is one of the world's largest energy companies, providing customers with fuel for transportation, energy for heat and light, and petrochemical products for everyday items. In June 2006, the firm announced it would establish the Energy Biosciences Institute. Then, in November 2006, it disclosed the formation of its biofuels business unit as part of its corporate strategy of identifying low carbon or renewable energy sources for the future. In 2008, BP began narrowing its investments in the clean-energy spectrum to those it considered commercially viable and a good match with its existing business. Biofuels made the cut, in part, because they fit into the firm's existing refineries, pipelines, and distribution networks.

The first area the SGI–BP alliance focused on was coal-bed methane as an alternative source of natural gas. SGI in collaboration with BP is striving to develop and use naturally occurring microbes to metabolize coal into methane, which can be harvested as natural gas. Coal-bed methane resources, globally and in the United States, are vast and relatively unexplored. Although not a renewable source of carbon, the organisms that convert carbon dioxide into methane can provide as much as tenfold improvement in comparison with the existing practice of mining and burning coal. Methane combustion, or burning, is also more environmentally friendly than coal combustion as a source of energy and produces a smaller global warming impact than coal combustion does. In announcing the project, Venter stated, "Through our research collaboration with BP, we will achieve a new and better understanding of the subsurface hydrocarbon bioconversion process which we are confident will yield substantial cleaner energy sources."[14]

Using naturally occurring microbes to metabolize coal into methane, which could be harvested as natural gas, SGI sought to apply commercial solutions to improve the recovery rates of natural gas, without genetically modifying bacteria or designing new microbes. In the first phase of the program, research centered on achieving a better understanding of the microbial communities in various hydrocarbon formations, such as coal, shale, petroleum, and natural gas. In identifying and describing naturally occurring organisms and their biological functions that thrive in these subsurface hydrocarbons, SGI used its expertise in DNA sequencing and microbial cell culturing to explore and understand microbial processes and produced the first comprehensive genomic study of the microbial populations living in these environments.

Studying the coal-bed biome was not as simple as studying more well-known bacterial systems, however, and SGI had to develop more sophisticated techniques to collect and analyze the subterranean organisms. BP and SGI worked together to identify and sample the appropriate subsurface hydrocarbon bases, technically, substrates, and SGI recovered the DNA from these samples and applied unique sequencing methods to find and analyze previously unknown DNA. Because SGI was not able to culture many of the underground organisms using traditional laboratory methods, SGI pioneered a new method of gathering and sequencing DNA from these organisms.

To gain access to these novel, uncultured microorganisms, SGI developed innovative microbial cultivation technologies and monitoring approaches. Moreover, SGI utilized a newly developed efficient DNA sequencing technique. This method, a complex process, required meticulous preparation and organization as well as cutting-edge equipment, including robots and high-throughput sequencing machines. The combination of new DNA sequencing and microbial cell culturing methods facilitated a new fundamental understanding of the dominant metabolic and chemical processes taking place in subsurface hydrocarbon microbial environments, allowing SGI and BP to continue their breakthrough efforts to develop coal-bed methane conversion technologies.

Nearly three years into the collaboration, Venter stated, "When we started [the BP program] we had no idea of the extent of the biology a mile deep in the Earth. It was stunning. We found the same density of organisms that we find in the ocean, all new types of organisms. We have hundreds of them that eat coal and coal substrates to produce different molecules. So we've been characterizing this biology now to see how scalable it would be to just change some of this biology and get a lot more natural gas out."[15]

Following the completion of the basic science research phase, the collaboration will undertake a series of field pilot studies of the most promising bioconversion approaches and seek to commercialize the technologies. The joint venture ultimately hopes to develop viable techniques for the bioconversion of subsurface hydrocarbons, such as coal, into cheaper, cleaner energy sources. In sum, instead of mining and burning coal, the SGI–BP venture seeks to generate more natural gas from coal thereby making the recovery of existing fuels, such as coal, more efficient and cost effective.

As part of the collaboration agreement, BP also made an equity investment in SGI, as noted earlier in this chapter.

Asiatic Centre Collaboration

In July 2007, SGI and the ACGT formed a multiyear research and development joint venture to develop high-yield, disease-resistant plant feedstocks.[16] ACGT is a wholly owned subsidiary of the Asiatic Development Bhd (now Genting Plantations Berhad), a publicly held oil palm plantation company, owning and operating plantations in Malaysia and Indonesia. AGCT focuses the application of genome technology to improve oil palm, among other crops. ACGT and Tan Sri Lim Thay, the chairman and CEO of Asiatic Development Bhd, made equity investments in SGI as part of the joint venture arrangement, as noted earlier in the chapter.

The partnership between SGI and ACGT focused on sequencing and analyzing oil seed plants, specifically oil palm and Jatropha, two of the most productive and promising oil-producing crops. Oil palm (*Elaeis guineensis*) is the world's highest-yielding oil seed crop. Malaysia and Indonesia account for about 80 percent of the globe's palm oil production and represent the main target markets for the joint venture's improvement efforts.

Jatropha (*Jatropha curcas*), a nonfood plant, represents one of the most promising bioenergy crops. It is a nondomesticated tropical tree with the potential to become one of the world's highest-yielding oil seed plant; its seed oil and biomass are ideal for biofuels production. Moreover, it can be grown in marginal, previously nonfood producing lands; so Jatropha does not compete with agriculture for food production. It is also an excellent candidate for enhancement by genetics and genetic engineering due to its very short generation time. Despite its germination period, however, Jatropha can be very productive for thirty to forty years.

In the project's first phase, SGI conducted in-depth analyses of the oil palm genome. By better understanding the oil palm's genome and breeding it with plants with useful traits, the collaboration hoped oil palm could become a better renewable energy source. The joint venture also developed diagnostic tests for plant diseases to enhance natural resistance mechanisms for the breeding and maintenance of disease-resistant energy crops, such as oil palm. The resulting genomic solutions will also help address ecological concerns with respect to the destruction of biodiversity through more efficient land use, higher agricultural yield, sustainable development, and improved stewardship of the plantation environment.

In May 2008, the joint venture completed the first draft assembly and annotation of the oil palm genome.[17] Unlocking the knowledge encoded in the genome helped increase the understanding of the oil palm plant and could lead to substantially improved oil yield. The oil palm genome consists of about 1.8 billion base pairs. SGI and ACGT sequenced a combination of two oil palm breeding stocks, tenera and dura, to generate a sevenfold coverage of the plant's genome. The draft genome yielded important information, such as the unique genetic variations linked to traits that differ in these two stocks. Because fruits with thinner kernel shells yield more oil, SGI and ACGT sought to understand the genetic basis for shell thickness. These molecular markers can be used in breeding and tissue culture-based approaches to address plant yield, oil quality, disease tolerance, and fertilizer requirements. The joint venture also continued to work on sequencing the Jatropha genome.

Then in May 2009, the joint venture completed a first draft of the Jatropha genome.[18] After the sequencing revealed that the Jatropha genome had about four hundred million base pairs, the respective teams turned to annotating the genome to identify specific genes of interest and to discover genetic variations for use in sophisticated, so-called marker-assisted, breeding. The two partners also explored the microbial life around the plant using environmental genomic techniques to sequence and analyze the Jatropha's root, soil as well as leaf bacterial and fungal communities. Understanding these environmental factors may help SGI and ACGT develop diagnostic tests for plant diseases and agents for disease control, resulting in healthier, more productive crops. These genomic solutions could also permit more efficient land usage, higher agricultural yield, and improved stewardship of the plantation environment. In the future, the joint venture hopes to use genomic methods to improve this oil seed crop to produce higher yielding plants for biofuels, biofertilizers, and disease-control agents. At present, it is unclear, however, whether SGI will successfully use synthetic genomic advances to increase Jatropha yield.

The collaboration will also evaluate microbes isolated from Jatropha roots for use as biofertilizers and disease-control agents. Studying soil microbes in the root zones of plants will assist in improving plant properties, reducing the use of fertilizers or chemical pesticides, and enhancing the disease resistance of plants.

ExxonMobil Collaboration

Representing a huge vote of confidence in Venter's work and the need to move away from feedstocks that compete with food production, such as corn, in July 2009, ExxonMobil Corp., one of the world's most sophisticated energy firms, announced a partnership with SGI. ExxonMobil plans to spend as much as $600 million, if research and development milestones are met, working with SGI on developing algae biofuels.[19] According to Venter, "Algae is the ultimate biological system using sunlight to capture and convert carbon dioxide into fuel."[20]

Representing a major signal to the biofuels industry, the collaboration sought to explore using algae to create biofuels commercially comparable with gasoline, diesel, and jet fuel. Combining Venter team's research prowess at genetic modification with ExxonMobil's financial heft and brawn marked a new chapter in the algae biofuels story. The $600 million investment is not large for a company of its size; however, ExxonMobil does not throw money around. After it carefully examined all available biofuel options, a two-year process, ExxonMobil zeroed in on algae, deeming it most likely to be produced economically on a sufficiently massive scale to compete in the world's energy mix. The firm then placed its bet on SGI to develop the technology.

SGI and EMRE established a comprehensive multiyear research and development strategic alliance focused on exploring the most efficient and cost-effective ways to produce the next generation of biofuels using algae. The plan is based on algae's potential to be an economically viable, low net carbon transportation fuel. As part of the agreement, SGI will receive milestone payments for developing biofuels products. Total funding by EMRE for SGI's research and development activities as well as milestone payments could amount to more than $300 million with the potential for SGI to gain additional income from licensing discoveries and products to third parties. The project also includes another $300 million in in-house spending at ExxonMobil, with possible future investments by the oil giant amounting to billions of dollars to build production facilities. On entering into the alliance, Venter stated, "This agreement between SGI and EMRE represents a comprehensive long-term research and development exploration program into the most efficient and most cost-effective organisms and methods to produce next-generation algae biofuel. We are confident that the combination of our respective expertise in science, research, engineering, and scale-up should unlock the power of algae as biological energy producers in methods and scale not previously explored."[21]

Photosynthetic algae, including microalgae (single-celled algae) and cyanobacteria (blue-green algae) use energy from sunlight to convert carbon dioxide into cellular oils (lipids) and certain types of long-chain hydrocarbons, with molecular structures similar to traditional crude oil, that can be further processed into fuels and chemicals, using existing refining techniques. Unlike hydrogen, fuels derived from algae (so-called "drop-in" fuels) are also compatible with existing transportation infrastructure, specifically gas stations. However, naturally occurring algae, even utilizing current methods including processes that resemble farming, do not carry out the processes at efficiencies or rates needed for the commercial production of biofuels. The venture, therefore, seeks to improve both aspects of algae biofuels production so as to make the technology viable and competitive in the global economy.

Under the agreement with EMRE, SGI, using its leading-edge scientific expertise and its proprietary tools and techniques in genomics, metagenomics, synthetic genomics, and genome engineering, will work to find, optimize, and/or engineer superior strains of algae. SGI will strive to design and synthesize algae with superior capabilities for converting carbon dioxide into traditional petroleum fuels. It will also design and develop the best production systems for the large-scale cultivation of algae and the conversion of algae products into useful biofuels, thereby playing a key role in bio-oil research and development.

ExxonMobil, using its scientific and engineering expertise, process development, scale-up, financial strength, and oil industry knowledge, as well as its immense financial strength, will develop various systems for increasing the scale of algae production, including determining in which type of system to grow the algae and how to manufacture finished fuels, such as gasoline, diesel, and jet fuel. ExxonMobil will also head development of process integration for commercial applications. Recognizing that it may take five to ten years from 2009, the initial date of the collaboration, to get production facilities running and require billions of dollars in additional investments to reach large-scale operations, as Venter noted, "The challenges are not minor for any of us, but I think we have the combined teams, scientific and engineering talents, to give this the best chance of success."[22]

SGI's work with ExxonMobil was not the firm's first step into the area of biofuels production. Prior to entering into the alliance with EMRE, SGI scientists had worked internally for several years to develop more efficient means to harvest the oils that photosynthetic algae produce. Previously, SGI had changed algae's gene structure to produce hydrocarbons similar to those that come out of the ground and tricked the organisms to pump out, instead of accumulate, the hydrocarbons as soon as they were produced. To do so, SGI engineered algal strains that secrete pure cellular oils (lipids) in a continuous process that is more efficient and cost effective than previously existing harvesting methods. These engineered algal cells secrete oils continuously through their cell walls, thereby facilitating the production of biofuels and chemicals in large-scale industrial operations. In the oil production process developed by SGI, algae secrete oil, which floats to the surface of the water. The oil is then skimmed off the water's surface and turned into biofuels and chemicals using refining techniques. It is hoped that a process similar to this will be used by ExxonMobil in the near future as an alternative source of petroleum products, but much work remains to be done.

The first step in the SGI–EMRE alliance will focus on identifying and developing the algae strains that are most productive and efficient at using energy from sunlight to convert carbon dioxide into oils. The alliance will then focus on altering the algal DNA to further boost the yield and alter the chemical composition of the oils produced, with the goal of making the oil as similar as possible to traditional petroleum. The alliance will also determine the best environment, or production system, for growing algae and producing biofuels, consisting of either an open pond or a closed container, called a bioreactor, or possibly both. After identifying the best production system for growing the enhanced algae strains, the focus will then turn to the development of large-scale production systems. The two firms will also co-develop the large integrated systems required for full-scale production, refining, and commercialization.

Algae as a Feedstock: Advantages and Disadvantages

Algae offer numerous advantages[23] in comparison to other biofuel feedstocks; however, a number of unknowns exist. Currently, researches are focused primarily on microalgae, or algal organisms that are less than 0.4 mm in diameter, believing that it will be possible to commercialize fuel production most effectively from these species. The advantages of microalgae come from its simple structure, its fast growth rate, its high oil content in certain species, and the ability to genetically engineer many of its beneficial properties. With proper genetic design, researchers believe microalgae will produce high amounts of oil similar in structure with crude oil at competitive costs. A similar oil structure is crucial, as the resultant algal biofuel chemistry will ensure that the fuels are compatible with existing transportation technology and infrastructure. Moreover, the crude oil similarly means that bio-oils from algae will also be used to manufacture a full range of fuels, including gasoline, diesel fuel, and jet fuel that meet the same specifications as existing products.

Importantly, algae yield greater volumes of biofuels per acre of production than other possible plant-based biofuel sources. For example, algae can currently yield more than 2,000 gallons of fuel per acre of production each year, compared with 650 gallons per acre annually for oil palm trees and 450 gallons per acre per year for sugarcane.[24]

Algae also do not require the use of current agricultural land, and play a minimal role in the food chain. Therefore, algae farms will not crowd out any agricultural land uses, and vast changes in the amount of algae on earth will have little to no impact on other organisms' food supplies or habitats.

Algae are also very simple to grow. Producing algae requires three inputs: sunlight and carbon dioxide for photosynthesis, and the algae themselves. Furthermore, algae do not require large amounts of fresh water; rather its cultivation uses low-quality nonpotable water, even wastewater and sewage effluents. It can even be grown in brackish water or farmed in seawater.

As photosynthetic organisms, algae consume large quantities of carbon dioxide, thereby creating a carbon-neutral cycle that removes greenhouse gases from the atmosphere, most of which are produced by the burning of current petroleum products. Because algae take up carbon dioxide to support their growth, combustion of the algae's oil creates no new carbon dioxide for the environment. In contrast, combustion of oil drilled from wells (or coal from mine) creates new carbon dioxide, because the drilled oil was not created from carbon dioxide and photosynthesis.

While clearly a potential competitor to conventional fuels, algae also have the potential to reduce carbon emissions from current conventional oil and coal uses. SGI and EMRE hope to engineer new strains of algae that can absorb large amounts of carbon dioxide emitted, for instance, by coal-fired power plants. Thus, algae farms could produce two income streams: one from selling algae oil to refineries and a second from capturing and reusing carbon dioxide.

Not all aspects of algae have been determined, and a number of unknowns still exist, however.[25] First, we do not know the impact of mass cultivation required for commercialization on genetically modified algae. To attain the requisite scale, it is uncertain whether algae will crowd one another out or emit toxic waste that halts the production process. Also, the oxygen produced during photosynthesis by the algae may result in the organisms poisoning themselves.

Second, if the algae are grown in open ponds so as to expose them to sunlight, undesirable organisms, among other contaminants, may mix with the algae colonies. A need exists to manage these potential pathogens which may reduce the biofuels yield.

Third, growing algae in open-pond systems requires vast amounts of land. Water losses must also be monitored and controlled, thereby increasing expenses. The most productive algae strains may not be the most physically fit strains, and natural algae strains, if allowed to contaminate the human-made strains, could consume the local resources to the point of killing their synthetic brethren. If this is the case, then special care necessitating containment in closed bioreactors will be required, further raising production costs, so that synthetic algae will not be out-competed by wild algae.

Fourth, extracting the oil from algae may be difficult. Full-scale production may require a significant number of processes, such as dewatering and the separation of the lipids, the oil, in the algae. Once the algae crop matures, in one to ten days depending on the strain, it would be harvested and oils would be separated from the aqueous slurry of algae and various major nutrients, such as nitrogen, phosphorus, potassium, and trace nutrients, such as iron. Large centrifuges are typically used in the separation process, representing a relatively high cost part of the production. In sum, it is uncertain whether algae will produce biofuels cheaply enough and not a sufficiently large scale to be financially viable in the near future.

The bottom line: will it be possible to produce genetically engineered biofuel using algae in commercial quantities? As Venter put it: "The question is scale. That the real bugaboo here for everybody.... But I think with the Exxon engineering team and their money, we have a chance to scale it up.... It is going to get down to the cost equation. And it's too early to know."[26]

More generally, a basic problem is that microorganisms, such as genetically engineered algae, want to take care of themselves. According to Venter, "If [the microbes] are unhappy with what they are doing, they are going to evolve way from what you want them to do. A key part of the future is going to be designing a system where they are not grossly unhappy with what they are doing."[27]

In the future, rather than genetically engineering algae, Venter wants to create a designer microbe, the heart of a biological engine, from scratch. In creating new fuels to replace petroleum and coal, Venter hopes to design the genome of a new organism that could produce fuels, among other products. A computer could direct a robot to chemically make the DNA strand encoding all the necessary information, and once constructed the new genome would be inserted into a bacterial cell. When activated, the chromosome would cause the cell to turn into the species that the scientist designed. The species could serve as a bioreactor making millions of copies of itself, with each copy producing fuel, perhaps even from sunlight with the carbon coming from carbon dioxide.

Synthetic Genomics is not alone in its quest to use new tools to make fuels, petrochemicals, food products, and pharmaceuticals. As analyzed in the next chapter, it faces at least four significant competitors that are much further along toward commercialization.

Notes

1. Quoted in Synthetic Genomics, Inc. (SGI), Press Release, "Synthetic Genomics, Inc. Launched to Develop New Approaches to Biological Energy," June 29, 2005.
2. SGI, "About Synthetic Genomics," http://www.syntheticgenomics.com/about/ (accessed September 13, 2009). In October 2010, SGI and the Venter Institute announced the formation of a new company, Synthetic Genomics Vaccines, that will focus on developing the next generation of vaccines. The new entity entered into a three-year collaboration with Novartis to use the Institute's research to speed the production of influenza strains needed to produce vaccines. J. Craig Venter Institute, Press Release, "Synthetic Genomics and J. Craig Venter Institute Form New Company, Synthetic Genomics Vaccines (SGVI), to Develop Next Generation of Vaccines," October 7, 2010.
3. SGI, Press Release, "Synthetic Genomics, Inc. Launched" and SGI, U.S. Securities and Exchange Commission (SEC) Form D, October 11, 2005. See

also Tricia Bishop, "Creating a Company Intended to Create Life," *Baltimore Sun*, June 30, 2005, A1; Barton Eckert, "New Biotech Firm Launches in Rockville," *Washington Business Journal*, June 29, 2005, http://washington. bizjournals.com/washington/stories/2005/06/27/daily24.html (accessed June 14, 2010); Antonio Regaldo, "Next Dream for Venter," *Wall Street Journal*, June 29, 2005, A1.

4. SGI, Press Release, "Dr. Aristides Patrinos Named President of Synthetic Genomics, Inc.," February 2, 2006.

5. For background on SGI, see SGI, "About Synthetic Genomics," http://www. syntheticgenomics.com/about (accessed September 13, 2009); SGI, "What We Do," http://www.syntheticgenomics.com/what (accessed August 14, 2009); SGI, *Corporate Overview*, n.d.; Life Science Analytics, Inc. Synthetic Genomics, Inc., Company Profile, April 24, 2009.

6. SGI, Frequently Asked Questions: Synthetic Genomics, http://www.syntheticgenomics.com/media/faq.html (accessed September 5, 2009).

7. J. Craig Venter, "The Richard Dimbleby Lecture: Dr J Craig Venter—A DNA-Driven World," December 4, 2007, http://www.bbc.co.uk/print/pressoffice/pressreleases/stories/200712_december/05/dimble (accessed February 24, 2010). For an analysis of how advances in genome engineering and design could affect the U.S. economy, see Bio Economic Research Associates, *Genome Synthesis and Design Futures: Implications for the U.S. Economy* (Cambridge, MA: Bio Economic Research Associates, 2007).

8. Michael S. Rosenwald, "J. Craig Venter's Next Little Thing," *Washington Post*, February 27, 2006, D1; Antonio Regaldo, "Next Dream for Venter," A1.

9. Quoted in Rosenwald, "J. Craig Venter's," D1.

10. SGI, Investors, http://www.syntheticgenomics.com/investors (accessed September 13, 2009); SGI, SEC Form D, October 11, 2005; Peter Marsh, "Leading the Evolution Out of the Fossil Fuel Age," *Financial Times* (London), October 22, 2007, 19. For a summary of SGI's financing transactions, see Global Markets Direct, Synthetic Genomics, Inc.—Alternative Energy—Deals and Alliances Profile, February 2010, GMDAE170465D, http://www.globalmarketsandcompanies.com (accessed April 14, 2010).

11. Ooi Tee Ching, "Lim Jatropha can be Nation's Next Big Crop," *New Straits Times Press* (Malaysia), August 6, 2008, 39; Goh Thean Eu and Ooi Tee Ching, "Asiatic Devt may Turn Acquirer," *New Straits Times Press* (Malaysia), July 12, 2007, 35.

12. Life Technologies Corp., Press Release, "Life Technologies Completes Investment in Synthetic Genomics, Inc.," June 2, 2010.

13. SGI, Press Release, "Synthetic Genomics, Inc. and BP to Explore Bioconversion of Hydrocarbons into Cleaner Fuels," June 13, 2007; SGI, *What We Do: Hydrocarbon Recovery and Conversion*, http://www.syntheticgenomics. com/what/hydrocarbonrecovery.html (accessed September 13, 2009); SGI, Frequently Asked Questions: BP and Synthetic Genomics, http://www. syntheticgenomics.com/media/bpfaq.html (accessed September 3, 2009). See also, Juan Enriquez, "The Future of Bioenergy," *Wall Street Journal*, October 12, 2007, A17; "BP Enters Two Biotech Collaborations: Feedstocks for Cellulosic Ethanol and Bioconversion of Subsurface Hydrocarbons," *Green*

Car Congress, June 13, 2007, http://www.greencarcongress.com/2007/06/bp_enters_two_b.html (accessed September 3, 2009); "BP Enters Two New Next-Generation Biofuels Ventures," *Ethanol & Biodiesel News* 19, no. 25 (June 18, 2007), http://ProQuest (accessed March 3, 2010).

14. Quoted in SGI, Press Release, "Synthetic Genomics Inc. and BP."

15. Interview, J. Craig Venter, "It Came from the Sea," *Wall Street Journal*, March 8, 2010, R6. See also Peter Marsh, "Leading the Evolution"; John A. Sullivan, "Scientist: New Lifeform Will be Engineered to Make Future Fuels," *Natural Gas Week* 24, no. 7 (February 18, 2008): 1, 15.

16. SGI, Press Release, "Synthetic Genomics Inc. and Asiatic Centre for Genome Technology Form Partnership to Sequence Oil Palm Genome," July 11, 2007. For details of the joint venture, see Proposed Joint Venture between Asiatic Centre for Genome Technology Sdn Bhd (Formerly Known as Cosmo-Lotus Sdn Bhd), A Wholly Owned Subsidiary of Asiatic Development Berhad and Synthetic Genomics, Inc., February 2, 2007; Ibid., May 15, 2007; Ibid., June 6, 2007. See also SGI, *What We Do: Agricultural Products: Improved Plant Feedstock*, http://www.syntheticgenomics.com/what/agriculture.html (accessed February 25, 2010).

17. SGI, Press Release, "First Draft of Oil Palm Genome Completed by Synthetic Genomics Inc. and Asiatic Centre for Genome Technology," May 21, 2008. See also Jenny Ng, "The Race for the Super Palm," *The Edge-Malaysia*, May 18, 2009, <BB 18.1.4>.

18. SGI, Press Release, "First Jatropha Genome Completed by Synthetic Genomes Inc. and Asiatic Centre for Genome Technologies," May 20, 2009.

19. SGI, Press Release, "Synthetic Genomics Inc. and ExxonMobil Research and Engineering Company Sign Exclusive, Multi-Year Agreement to Develop Next Generation Biofuels Using Photosynthetic Algae," July 14, 2009; SGI, *What We Do: Next Generation Fuels & Chemicals*, http://www.syntheticgenomics.com/what/renewablefuels.html (accessed September 13, 2009); SGI, *Next Generation Algae Biofuels Fact Sheet*, http://www.syntheticgenomics.com/media/emrefact.html (accessed February 25, 2010). See also Jad Mouawab, "Exxon to Invest Millions to Make Fuel from Algae," *New York Times*, July 15, 2009, B1; Thomas Kupper, "Deal Blooms for Algae Biofuels Research," *San Diego Union-Tribune*, July 15, 2009, A1; Guy Chazan, "Producers—Bio Oil Looks to Biofuels," *Wall Street Journal*, October 19, 2009, R2; Russell Gold, "In Strategy Shift, Exxon Plans $600 Million Biofuels Venture," *Wall Street Journal*, July 15, 2009, B4; Catherine Brahic, "Renewable Oilman," *New Scientist* 23, no. 2718 (July 25, 2009): 25; Angel Gonzalez, "Exxon, Biotech Pioneer to Develop Algae Fuel," *Globe and Mail* (Canada), July 15, 2009, B5; Kristen Hays, "Exxon Mobil Follows the Biofuels Trend," *Houston Chronicle*, July 15, 2009, Business Section, 1.

20. Quoted in Mouawad, "Exxon to Invest Millions."

21. Quoted in SGI, Press Release, "Synthetic Genomics Inc and ExxonMobil Research and Engineering Company."

22. Quoted in Gold, "In Strategy Shift."

23. See Carol Freedenthal, "Liquid Fuels from Algae Show Many Advantages," *Pipeline & Gas Journal* 237, no. 1 (January 1, 2010): 18–19; Tim Studt,

"Algae Promises Biofuel Solutions," *Laboratory Equipment* 46, no. 12 (March 2010): 1, 8–9.

24. SGI, Next Generation Algal Biofuels Fact Sheet.

25. Andres F. Clarens et al., "Environmental Life Cycle Comparison of Algae to Other Bioenergy Feedstocks," *Environmental Science and Technology* 44, no. 5 (January 2010): 1813–919, concluded that conventional crops have lower environmental impacts than algae in energy use, greenhouse gases, and water. For a rebuttal, claiming the study used thirty-year-old data, see Algal Biomass Organization, Press Release, "Algal Biomass Organization Questions Accuracy of University of Virginia Algae Life Cycle Study," January 25, 2010.

26. Interview, J. Craig Venter, "It Came From the Sea."

27. Quoted in Jeff Tollefson, "Energy: Not Your Father's Biofuels," *Nature News* 451 (February 20, 2008), doi:1038/451880a, http://www.nature.com/news/2008/080220/full451880a.html (accessed September 29, 2009).

8

Competitors in the Race to Commercialize Biofuels and Chemicals

In the race to commercialize biofuels and chemicals, SGI faces a number of competitors that use one of two feedstocks: algae or sugarcane. This chapter profiles four of these companies: Sapphire Energy (Sapphire), Solazyme, Amyris Biotechnologies, and LS9. Similar to SGI, the first two firms seek to perfect and scale-up algae-to-fuel processes. Amyris Biotechnologies and LS9 use sugarcane as their feedstock with Solazyme utilizing sugar to accelerate its algae-production process. These four firms are building (or have built) pilot plants and/or commercial facilities with Sapphire, Solazyme, and Amyris Biotechnologies on the verge of commercialization. To date, SGI does not have a pilot plant.

Sapphire Energy, Inc.

Sapphire, founded in May 2007, uses algae to produce petroleum.[1] Like SGI, Sapphire's cultivation of genetically modified algae does not require crop or farmland or use potable water.

Background

A group of friends—entrepreneur—scientist, Jason Pylle, MD, Ph.D., Kristina Burow, a chemist-turned venture capitalist and a principal with ARCH Venture Partners, and Nathaniel David, a biologist and company builder—talked about the perils of ethanol and discussed which biological organisms might be best suited as a feedstock for biofuels. They found a large number of academic papers on algae written by Stephen Mayfield, Ph.D., then a professor at the Scripps Research Institute in San Diego, now the director of San Diego Center for Algae Biotechnology at the UCSD where he is also a faculty member. After speaking with Mayfield, they settled on algae, one of the world's fastest growing plants, and set up their company in San Diego. The new firm snapped up patents, erecting a fence around its proprietary technology to engineer algae to produce oils with molecules similar to crude oil. After preliminary work focusing on algae's potential, in 2008 they announced Sapphire's existence and its tens of millions of dollars in venture capital funding.

Sapphire's Successes

Starting with about forty strains of oil-producing algae, the firm genetically engineered some one hundred thousand strains, to isolate their strains that create lots of oil very quickly. The strains faced various tests designed to determine: their hardiness; ability to flourish outdoors in saltwater; resistance to predator attacks; ease at harvesting; and lipid content. Modifying the genome of various algae strains and moving genes into algae to force it to make the specific chemical building blocks needed to produce biofuels, by September 2007, the firm found a winner in algae whose oil is molecularly similar to light "sweet" crude petroleum. In partnership with the University of Tulsa, Department of Chemical Engineering, the firm developed a strain that did not escape into the wild and a method of keeping wild algae from invading its cultivation and damaging production.

The company unveiled the world's first renewable gasoline in May 2008.[2] Its "green" crude has the same energy density as gasoline. Because it is compatible with the existing petroleum infrastructure, as a "drop-in" fuel, it can be shipped in existing pipelines and refined the same way as gasoline is. It is low in carbon, very low in sulfur, and contains no benzene.

The company tested its algae-based jet fuel successfully on a Continental Airlines commercial two-engine jet flight in January 2009.[3] The flight operated with a biofuel blend, consisting of a 50 percent biologically derived fuel, including components derived from algae (provided by Sapphire) and Jatropha (provided by another company), and 50 percent traditional jet fuel in the number 2 engine. The aircraft's number 1 engine operated on 100 percent traditional jet fuel. The plane's number 2 engine burned some 4 percent less fuel than engine number 1; using less fuel translates into substantial dollar savings as well as environment and energy security benefits.

Funding

By mid-September 2008, just prior to the stock market meltdown, Sapphire had raised substantially more than $100 million in two equity rounds from venture capital firms, including ARCH Venture Partners, Venrock, and a medical research charity, Britain's The Wellcome Trust, a funder of the Human Genome Project.[4] As noted earlier, Kristina Burow, a principal at ARCH Venture Partners, cofounded Sapphire. Bill Gates' investment holding company, Cascade Investments LLC, participated in the second round of venture capital financing.

The initial three institutional investors gave Sapphire an open checkbook, not based on the usual venture model of set rounds, milestones, and valuations. The firm could draw as much capital as needed to commercialize its technology as rapidly as possible. The firm possessed the ability to build its first commercial facility with venture capital and federal government funding.

The firm is currently building a test facility outside Columbus, Luna County, New Mexico, with ponds that tap water from an underground aquifer. The integrated algal biorefinery project will demonstrate the validity of Sapphire's process that cultivates algae in ponds for conversion into biofuels. The facility was funded, in part, by $104.5 million in federal monies, a $50-million grant by the DoE through its Integrated Biorefinery Program and $54.5 million in the form of loan guarantees through the USDA's Biorefinery Assistance Program.[5] The Energy Department funds came from the 2009 American Recovery and Reinvestment Act. Projects selected under this program would assist in validating refining technologies, help lay the foundation for commercial-scale development of a U.S. biomass industry and were part of an ongoing effort to reduce American dependence in foreign oil, spur the creation of a domestic biofuels industry, and provide new jobs in rural areas. The non-ARRA loan guarantee program was authorized by a provision in the 2008 Farm Act to promote the development of new technologies for fuel production from noncorn kernel starch biomass sources.

Sapphire has an option to purchase 2,200 acres (and water rights) near Columbus where it plans to build more than 200 acres of ponds to cultivate algae capable of producing one million gallons of fuel annually. By 2011, with the facility operative, the company expects to turn out one million gallons of biodiesel and biojet fuel a year. Then, by 2018, with an expanded facility, the firm hopes to produce one hundred million gallons per year and by 2025, one billion gallons annually.[6]

Sapphire's Technological Breakthroughs

Sapphire has pioneered in two breakthroughs in its algae-to-fuel process. First, at present it takes considerable of energy to separate the oil produced by the algae from the water in which the organisms grow, thereby defeating, in part, the purpose of using algae. Sapphire has designed a cost-effective dewatering process to separate the water from the oil.

Second, to harvest the oil the algae produce requires extracting them from their ponds, drying them out, and breaking open their cells. This is a tedious process. Sapphire uses a proprietary process to more efficiently extract the oil. The firm then subjects the oil to catalytic cracking to obtain gasoline or hydrogen treatment to obtain diesel or jet fuel.

Only time will tell whether it will be possible for Sapphire to scale-up production at its New Mexico facility; however, like SGI, Sapphire is placing its bet on genetically modified algae.

Solazyme, Inc.

Solazyme, Inc. (Solazyme), founded in March 2003, by Harrison F. Dillon, Ph.D., its current president and chief technology officer, and Jonathan S. Wolfson, its current CEO, is an algal biotechnology company. It genetically engineers microalgae to produce fuels and specialty chemicals.[7] Viewing itself as a renewable oil-production company, not merely a biofuels firm, led to the formation of its health services business unit that makes personal care products, nutraceuticals (a foodstuff providing health benefits), and cosmetics, using its core technology.

Solazyme's Successes

In pursuing its business development path, Solazyme has taken lead in transferring its technology to commercial production. Since the spring of 2007, it has produced its oil-based fuel, Soladiesel, using standard industrial fermentation facilities with contract manufacturing partners in Pennsylvania and California. As it looks to expand in its production capabilities, the firm can perform its novel conversion technology process in commercial-scale facilities already built to serve global markets. In addition to jet fuel, the firm sought to develop diesel fuels because industries using both fuels will not be able to replace them anytime soon with hybrid or all-electric plug-in engines.

Its two types of Soladiesel, RD and BD, were the first algal-derived fuels to have been successfully road-tested for thousands of miles in a variety of unmodified diesel vehicles. In June 2008, its Soladiesel RD passed American Society for Testing and Materials (ASTM) D-975 specifications.[8] Its Soladiesel BD also meets the ASTM D-675 specifications.[9] Both types of Soladiesel, as "drop-in" fuels, are compatible with existing petroleum industry infrastructure, refineries, pipelines, gas stations, existing vehicles on the road, and can be used in factory-standard diesel engines, without modifications. The firm's Soladiesel also produce about a 90 percent reduction in greenhouse gas emissions and a 30 percent decrease in particulate emissions in comparison to today's petroleum-based ultra-low-sulfur diesel.[10]

In September 2008, the firm produced the world's first microbial-derived jet fuel, SolaFuelJet. The fuel passed all eleven specifications for aviation jet fuel.[11]

One year later, in September 2009, the U.S. Department of Defense selected Solazyme to provide 1,500 gallons of algae-derived jet fuel from the firm's commercial-scale facilities for testing and certification by the U.S. Navy. Successfully tested in October 2010, the Navy contract enabled the company to demonstrate that its jet fuel met all military specifications and functional requirements.[12] By delinking U.S. military bases, ships, and planes from the global petroleum supply chain so that if events cause the price of petroleum to skyrocket and supplies to become unavailable, the American military will be able to meet its needs and attain energy security. The military-driven project has civilian benefits. The military's substantial buying power may help spur the expansion of algae-based fuels into the civilian market.

In an effort to move to energy independence and spur the use of alternative fuels, also in September 2009, the U.S. Department of Defense awarded the firm an $8.5 million contract to research, develop, and demonstrate the large-scale production of algae-derived low-carbon biofuels to meet the U.S. Navy's specifications for its ships.[13] Solazyme used its technology to provide Soladiesel F-76 Naval Distillate fuel for testing and certification to demonstrate that it met all military specifications and functional requirements. F-76 is similar to the diesel fuel that is the primary shipboard fuel used by the U.S. Navy. Together with an R&D component, the award called for the delivery of more than twenty thousand gallons of 100 percent algae-based fuel for testing and certification.

Scientific Background

Based on its experience, unlike SGI and Sapphire, Solazyme decided not to make fuels directly from microorganisms, because the firm believed that process would be too expensive. In making a hydrocarbon fuel directly from a microorganism, in its view, the process also tends to be toxic, reducing the efficiency of the fermentation system.[14]

The process used by Solazyme combines genetically modified strains of algae with an uncommon approach to their growth to reduce the cost of making biofuel. Solazyme developed a fermentation platform rather than growing algae in ponds or enclosed in plastic tubes exposed to the sun, where, in other systems, they make their own sugar. The firm grows algae inside of huge, dark stainless steel industrial fermenters—containers. To provide energy to the algae in the fermentation process, Solazyme feeds sugar to algae that helps them generate oil. Keeping the algae in the dark also causes them to produce more oil, the company found, than if they were exposed to the light. While their photosynthetic processes are inactive, other metabolic functions that do the conversion work become active, enabling the organisms to produce oils more efficiently and making the oil recovery less costly. Feeding algae sugar, among other cellulosic feedstocks, makes it possible to grow them in much higher concentration levels than if grown in ponds using energy from the sun. Although sugar provides a concentrated energy source, this step in the process is more expensive because it requires feeding the algae sugar, not using free sunlight and carbon dioxide. However, these higher energy concentrations reduce the amount of infrastructure needed to grow the algae and make it easier to extract the oil, when the algae are fat and grown thereby reducing the overall production cost. After being squeezed out, the oil is processed using standard refining technology to make a range of fuels, including diesel and jet fuel. Inventive steps are not needed on the infrastructure side of things. The firm accepts a modest refining cost rather than focusing on making new biofuels that will require multiyear regulatory reviews costing tens of millions of dollars and require new infrastructure to be built prior to commercialization.

The firm uses various strains of algae to produce different types of algal-oil. Some algae produce triglycerides similar to those produced by soybeans, among other oil-rich crops that are plentiful in unsaturated fats, contain no transfats, and are low in saturated fats. Others produce a mix of hydrocarbons similar to light crude petroleum.

Although its fuels can be transported and delivered using existing fuel infrastructure and used in unmodified vehicles, at present, Solazyme's method of creating fuel is not currently cost effective to compete with petroleum. To find a cheaper biomass to reduce costs, in May 2009, Solazyme announced an agreement to test sugars produced by BlueFire Ethanol Fuels, Inc. (BlueFire), now renamed BlueFire Renewables, Inc.[15] BlueFire had developed a patented process for making sugar from garbage, for example, at landfills. Using the BlueFire process will enable Solazyme to decrease its costs. Making biofuels from various cellulosic sources relies on processing them to create simple sugars that may, however, contain poisonous substances such as lignin, an amporphous polymer that provides rigidity and together with cellulose forms the woody cell walls of plants. Unlike other organisms, algae tolerate lignin, thereby skipping the need to separate the lignin from sugar.

In addition to optimizing the algae and striving to reduce its costs, the firm improved the amount of oil that the algae produce. It has achieved high productivity, with algal cells that are over 75 percent oil by dry weight within oil vesicles, the membranous, fluid-filled pouches, producing hundreds of grams of oil per liter of cells, and making extraction less difficult.[16]

Funding

To gain funding, Solazyme turned to venture capital firms and private investors raising some $130 million, to date, from these sources. In December 2005 and June 2006, it completed a $3 million round of equity financing, led by The Roda Group.[17] In February and April 2007, the firm raised an additional $8.66 million in a second round of equity financing from The Roda Group, Harris & Harris Group, Inc., and other individual and institutional investors.[18] BlueCrest Capital Finance, L.P. provided $5 million in debt financing in September 2007 that was repaid in May 2010.[19]

In a third equity financing round, completed in February 2009, the company garnered some $57 million, followed by a fourth round, completed in August 2010, that raised $59.7 million.[20] Braemar Energy Ventures and Lightspeed Venture Partners, joined by other investors, including VantagePoint Venture Partners and its major existing investors, including The Roda Group and Harris & Harris Group, provided the third round of financing. The new funding, the bulk of which came from VantagePoint Venture Partners, enabled Solazyme to boost its production capabilities and begin to attract customers, initially the U.S. Navy.

Solazyme has also received funds from federal government grants and an oil giant, Chevron. In September 2007, it received a $2 million Advanced Technology Program Award from the U.S. National Institute of Standards and Technology to develop a biopetroleum derived from marine microorganisms.[21] The grant helped the firm advance the achievement of major technical milestones toward the development of biopetroleum that is compatible with the existing petroleum industry infrastructure. Then, in January 2008, the firm signed a biodiesel feedstock development and testing agreement with Chevron Technology Ventures (CTV), a division of Chevron U.S.A. Inc.[22] The relationship with CTV represented an important step toward the commercialization of Solazyme's technology that fits into Chevron's existing refining and fuels distribution infrastructure. Like SGI and ExxonMobil, Solazyme also sought to take advantage of Chevron's know-how in getting fuel produced in bulk and distributed to consumers.

As the federal government sought to accelerate the construction and operation of biorefineries in the United States, in December 2009, the firm received $21.8 million grant from the DoE through the Integrated Biorefinery Program.[23] With capital markets then unable to supply the tens of millions in funds to get the production of its algae-based biofuels off the ground, the grant enabled Solazyme to build its first integrated biorefinery in Riverside, Pennsylvania. By validating the projected economies of a commercial-scale biorefinery producing various biofuels from algae, the grant marked a major step toward the commercial-scale production in a cost-effective manner of algae-based fuel. A key factor in the company's decision to locate its integrated biorefinery where it did was that the fermenters on-site required minimal conversion and upgrading.

Chemical Applications

Recognizing the utility of its process beyond biofuels, in March 2010, Solazyme signed a research and development agreement with Unilever to develop oil derived from algae for use in personal care products, including soaps. The agreement followed a yearlong collaboration between the two firms, in which Unilever successfully tested Solazyme's renewable algal oils in Unilever's product formulations.[24]

Today, the biggest obstacle to commercialization of algae-based products centers on scaling up to the size needed for mass use, for instance, as a fossil-fuel replacement, on the order of tens of billions of gallons per year and significantly lowering production costs. Seemingly, Solazyme's fermentation process is one of the closest, if not the closest, to a mature system. The firm may have the lead in solving the scale-up question thereby overcoming the risks involving ramping up production.

In May 2011, Solazyme raised some $180 million in an initial public offering. The company plans to use these funds to accelerate its business strategy by entering into feedstock arrangements and establishing manufacturing capacity with partners.

Apart from algae, sugarcane is a favorite with the biofuels industry because its efficient form of photosynthesis enables it to grow quickly. The next two companies, Amyris Biotechnologies, Inc. (Amyris) and LS9, Inc., use synthetic biology to produce products through the fermentation of plant-based sugars, among other feedstocks, by altering the metabolic pathways of microorganisms, such as yeast and bacteria. Researchers create living factories that can ferment plant-based sugars into hydrocarbon molecules found in traditional petroleum fuels, among other products.

Amyris Biotechnologies, Inc.

Amyris was founded in 2003 by Jay D. Keasling, Ph.D. and three post-doctoral fellows from his laboratory, Neil Renninger, Ph.D., Kinkead Reiling, Ph.D., and Jack D. Newman, Ph.D. Keasling, a professor of chemical engineering and bioengineering at the University of California, Berkeley (U.C. Berkeley), today serves as the head of the company's Scientific Advisory Board.[25]

Amyris began as a company working on the anti-malarial drug, artemisinin. It then transformed itself into a renewable products company using its cutting-edge synthetic biology technology platform to modify the metabolic pathways of microorganisms, yeast strains, to engineer factories that transform sugar into different molecules having a wide variety of practical pharmaceutical, energy, and chemical applications. The microbes chew up broken plant material, such as sugarcane, and convert it to specialty chemicals and transportation fuel. Busting up the plant matter is called pretreatment and is done with acid or heat and pressure. The entire system is called consolidated bioprocessing with all of the conversion processes done by one microorganism.

Amyris proved its technology platform in an earlier nonprofit project to reduce the production cost of artemisinin, an anti-malarial drug. Keasling's team identified the genetic pathway and developed microbial systems for producing a precursor of artemisinin. Using this technology platform, Amyris developed capabilities to produce specialty chemicals and high performance, renewable hydrocarbon transportation biofuels that are compatible with current engines and distribution infrastructures. The firm applies synthetic biology tools to produce a broad range of products, by combining its technology with production capabilities in Brazil through its subsidiary, Amyris Brasil SA. Amyris seeks to join technology, production, and distribution so as to commercialize and scale-up its products effectively.

Pharmaceutical Background

Amyris is famous for applying synthetic biology technology to manufacture an anti-malarial drug, artemisinin, naturally extracted from a plant, sweet wormwood (*Artesmisia annua*) that grows in China and Vietnam. Although the drug in its natural form is perhaps the most effective weapon against malaria, it is expensive, labor intensive, and time consuming to produce, thereby rendering it in short supply, hampered also by challenging climate and agricultural conditions. Its purification process, using diesel fuel, to extract the active ingredients may result in the final product retaining toxic impurities.

As a nonprofit project, initially, Keasling's laboratory at the U.C. Berkeley pioneered in perfecting techniques that rewrote the metabolism of microorganisms. By modifying the structure of a microorganism's genetic structure, thereby creating genetically modified microbes, Keasling designed microbial chemical factories that could produce a variety of drugs, chemicals, and biofuels. His team re-engineered brewer's yeast to make a low-cost anti-malarial drug by creating a biochemical pathway enabling microbes to make the precursor that could easily be converted into the drug.

By making about fifty genetic changes, Keasling and his colleagues tweaked a pathway to persuade brewer's yeast (*Saccharamyces cerevisiae*) to produce and secrete large amounts of artemisinic acid, a compound just a few chemical steps away from artemisinin.[26] After discovering the genes required for the production of artemisinic acid, the team functionally expressed them in a modified yeast strain designed to produce the artemisinin precursor artemisinic acid.[27] The team used sophisticated bioinformatics techniques and a synthetic biology screen to discover the necessary genes in a modified yeast strain derived from brewer's yeast. The process utilizes made-to-order yeast, rather than *E. coli*, a bacterium, because Keasling's team found that yeast is more scalable to commercial operations.

To move the project ahead, Keasling teamed up with the Institute for OneWorld Health (OneWorld), the first nonprofit pharmaceutical company in the United States, and Amyris, the company Keasling co-founded, in a three-way partnership to address the critical need in the developing world for a cost-effective malaria treatment. The tripartite collaboration, formed in late 2004, was funded by a $42.6 million five-year grant from the Bill and Melinda Gates Foundation to OneWorld to develop a low-cost process for the production of artemisinin.[28] The Gates Foundation grant provided: $8 million to Keasling's laboratory to complete the research work; $12.2 million to Amyris to improve the output of the microbes and decrease the cost; and $22.4 million to OneWorld to do the testing needed to satisfy regulators that the synthetic drug was equivalent to the plant-derived product.

After receiving the Gates grant, the three collaborators divided up the work. Keasling's laboratory at U.C. Berkeley conducted the basic research to perfect the microbial factory for artemisinin. Amyris developed and scaled up the process for industrial fermentation and commercialization. It developed a business scale process for the production of the drug, coupling microbial production systems with synthetic biochemistry. OneWorld implemented a global regulatory and access strategy for the drug by performing the nonchemical regulatory work required to allow the low-cost, microbially based product to be substituted for the plant-based item.

After the U.C. Berkeley issued a royalty-free license in 2006 to Amyris and OneWorld to develop the artemisinin technology and ensure affordability,[29] in March 2008, the OneWorld, Amyris, and sanofi-aventis, a leading pharmaceutical company, entered into an agreement for the development of semi-synthetic artemisinin.[30] Under the agreement, the parties worked jointly to develop and design pilot and commercial-scale manufacturing processes for artemisinin. In licensing its work using a microbe to create artemisinin, Amyris provided its engineering expertise using synthetic biology tools; sanofi-aventis contributed its fermentation and chemistral process development expertise; and OneWorld focused on achieving public policy and access goals. Because sanofi-aventis ran into unexpected production obstacles, it plans to manufacture the drug at close to the average price of the plant-derived version.

Biofuel and Chemical Applications

Having developed the technology platform to engineer microbes to make the chemical precursor of artemisinin, in January 2006, Amyris' founders began to consider seriously what other compounds their bioengineered microorganisms might produce. They asked: what would a perfect fuel look like if developed from scratch? Prodded by several leading venture capitalists, they decided to alter the metabolic pathways of microorganisms to engineer living factories to transform sugarcane to create a biofuel that would have a higher energy content than ethanol, would be usable in current auto, diesel, and jet engines, and be insoluble in water so it could be transported through the same pipelines that move petroleum.[31]

They quickly realized that ethanol is not the best biofuel. In high concentrations, it cannot be delivered through existing pipelines and pumps. It must be carried in its own tanks and pipelines. It contains 30 percent less energy than gasoline and it must be mixed with gasoline before being burned in conventional engines. At the end of ethanol fermentation process, the mixture must be distilled to separate ethanol from water.

Amyris' founders concluded that the firm could engineer metabolic pathways in a microbe, yeast, to produce a target molecule, farnesene that could be used as a chemical ingredient for consumer and industrial products and transportation fuels, via the fermentation of a feedstock, sugarcane. Farnesene, a pleasant-smelling oil accounts for part of odor of apple skins.

Today, Amyris uses synthetic biology technology to reprogram yeast to function as living factories for the production of fuels and chemicals. The firm possesses the ability to perform finishing processes to convert farnesene into a variety of end products. With respect to biofuels, the final product, a "drop in" hydrocarbon fuel, is produced by hydrogenating farnesene (combining it with hydrocarbon) into farnesane, a highly combustible power source with properties similar to those of diesel fuel.

Amyris refers to its products as "no comprise" because it designs them to perform equal to or better than currently available products.

For example, Amyris' renewable diesel fuel has virtually no sulfur and creates less carbon monoxide and hydrocarbon-exhaust emissions than fossil fuels without compromising engine performance. Its biodiesel fuel can be blended with petroleum-based diesel fuel at up to a 50 percent concentration rate, far higher than the existing 10–20 percent rate for existing biodiesel fuels. It can withstand extremely low temperatures, does not clog filters, and can be stored for long time periods with degradation. Another key attribute of its diesel fuel is that it is a hydrocarbon, the same component found in current petroleum fuels, enabling it to be used in any type of diesel engine. It can also be easily distributed using the existing fuel infrastructure.

The firm's future commercial prospects depend, however, on its ability to produce these target molecules from feedstock at high volumes. Its production costs are mainly driven by its ability to increase the yield derived from its yeast strains, among other factors. The term "yield" refers to the amount of the desired target molecule produced by a fixed amount of feedstock. To successfully enter the transportation fuels and chemicals market, its yeast strains need to produce end products at substantially higher yields than it has achieved to date, thereby lowering its costs.[32]

Funding

After deciding to get into biofuels and specialty chemicals, Amyris quickly obtained some $55.6 million in two rounds of venture capital funding in 2007–2008.[33] The company raised $14.1 million in its first round of venture capital equity funding in April and May 2007 from three leading firms. The financing was led by Khosla Ventures II, L.P., with participation from Kleiner Perkins Caufield & Byers (KPCB Holdings, Inc.) and TPG Biotechnology Partners II, L.P. In addition to helping it create the low-cost anti-malaria drug, the financing enabled Amyris to add to a new program aimed at producing renewable, high-performance biofuels with increased cost effectiveness. Then, in September 2007 and April 2008, Amyris raised an additional $41.5 million in a second round of equity funding. Duff Ackerman & Goodrich Ventures led the financing, joined by existing investors, Khosla Ventures, Kleiner Perkins Caufield & Byers, and TPG Biotechnology. Amyris used these funds to further the development of and scale-up its technology for the production of three transportation biofuels, gasoline, diesel, and jet, and to support business initiatives enabling the firm to bring its biofuels to market.

Despite the severe economic downturn in 2008–2009, funds continued to pour in. While embarking on a series of agreements with various Brazilian firms, discussed next, in the summer and fall 2008 and January 2009, Amyris obtained $66.1 million in a third round of equity financing.[34] Investors included GrupoCornélio Brennand de Brasil and Naxon UK, as well as Khosla Ventures, Kleiner Perkins Canfield & Byers, TPG Biotechnologies, and Votorantim Novos Negócios. In October 2008, in the midst of the stock market meltdown, Votorantim Novos Negócios, the venture capital and private equity arm of the Votorantim Group, a privately held Brazilian industrial company, made an equity investment of some $30 million in Amyris.[35] In a fourth equity round, Amyris secured approximately $61.1 million in the summer and fall of 2009 and January 2010. In April 2010, Amyris announced that Singapore-based Temasek Holdings invested $47.8 million in the company, bringing to a total of nearly $250 million in private funding and in June 2010 Total SA invested $133 million, bringing the private funding to an aggregate of more than $380 million.[36]

Ramping-up Biofuel and Chemical Production

With funding in hand, beginning in 2008, Amyris turned to commercialization of its biofuels and chemicals. Looking to Brazil's world leadership position in alternative fuel production and the country's established infrastructure, the firm entered into a series of agreements to strengthen its position in Brazil, the site of the lowest cost feedstock available, sugarcane, to make its products. In March 2008, Amyris established Amyris–Crystalsev Pesquisae Desenvolvimento de Biocombustiveis, a Brazilian joint venture between Amryis and Crystalsev Comercio e Representacao Ltd. (Crystalsev), to work with Brazilian sugarcane mills and fuel producers to gain access to sugarcane feedstock and to scale-up the production of Amyris' diesel fuel and chemicals.[37] A pioneer in the commercialization of products made from sugarcane, Crystalsev is one of Brazil's largest ethanol distributors and marketers. Amyris holds the majority stake in the joint venture; Crystalsev, the remaining interest. Crystalsev contributed its commercial expertise to the venture. Santelisa Vale, Brazil's second largest ethanol and sugar producer and the majority owner of Crystalsev (prior to the sale of its stake to a French commodities group in 2009), agreed to provide two million tons of sugarcane crushing capacity and adapt the new technology starting at its flagship mill, rather than the venture building a new facility. Santelisa Vale also provided technical and engineering expertise to accelerate the development and commercialization of Amyris' biodiesel.

Along with pushing forward in 2009 in Brazil, the firm also focused on its efforts in the United States. Amyris opened a small pilot plant in Emeryville, California, in November 2008 to produce sufficient quantities of biodiesel fuel for testing and to secure certification from the U.S. Environmental Protection Agency (EPA).[38] The U.S. plant served as the technical gateway to commercialization in Brazil. It demonstrated the firm's technology in scaled-down process equipment that was representative of full-scale commercial operations. The pilot plant also generated essential engineering data for designing the firm's full-scale plants and generated product samples for performance testing.

In April 2009, Amyris registered its diesel fuel with the EPA.[39] It marked the first time a hydrocarbon-based fuel made from a renewable plant-derived feedstock was registered for commercial sale in the United States. By registering its diesel fuel with the EPA, Amyris could use the fuel in demonstrations by a leading engine manufacturer at up to 20 percent (subsequently raised to 35 percent) blends and have it road-tested in cars and trucks.

Continuing to look south to Brazil, in addition to building a pilot plant in Campinas, Brazil, designed to validate its technology, the firm opened the Amyris Renewable Products Demonstration Facility in Campinas in June 2009 to demonstrate scale operations of its fuels and chemical manufacturing processes.[40] With this demonstration facility, Amyris achieved fully integrated capabilities to move its technology from the laboratory to pilot to demonstration and ultimately to commercial scale, continually testing the results throughout the chain. The demonstration plant allowed the firm to validate its commercial equipment design and manufacturing processes and produce more than ten thousand gallons of its products under conditions replicating full-scale manufacturing.

Amyris moved rapidly to commercialize production in Brazil starting in 2011, beginning with mills it owns or controls. To implement this goal, in December 2009, Amyris entered into an agreement with São Martinho Group to acquire a 40 percent stake in the modern, efficient, ethanol-producing Bioa Vista sugar mill, owned and operated by the Brazilian firm. The parties began working to convert this mill to produce Amyris specialty chemicals and diesel fuels.[41] Then, in February 2010, the Stratus Group invested $5.4 million in Amyris Brasil SA.[42] These funds were applied to the $75 million investment

Amyris would need to make in the Bioa Vista sugar mill. Amyris Brasil SA, its Brazilian subsidiary, oversees the company's scale-up, production, and distribution in Brazil.

To further the firm's commercialization efforts in Brazil, in December 2009, Amyris Brasil announced that it had entered into nonbinding letter of intent agreements with three Brazilian sugar and ethanol producers to provide additional access to sugarcane feedstock.[43] As part of its business plan, Amyris intends to build its production in Brazil through agreements under which the firm provides technology and plant design and the mill owners contribute capital to convert their mills to produce fuels and chemicals in return for a share of the expected higher gross margins on these products. Amyris Brasil will distribute and market these items to end customers. These so-called "capital light" agreements with three Brazilian firms, Bunge Ltd., The Cosan Group, and Açucar Guarani, represented key steps by Amyris toward building a fully integrated biofuels and specialty chemicals company, encompassing the requisite technology, industrial-scale manufacturing, and product distribution capabilities, at least in Brazil by working with established Brazilian companies to build new, so-called "bolt-on" facilities adjacent to their existing mills. However, the mill owners may back out of contracts if they are unwilling to contribute capital, thereby limiting the production of Amyris' products or making them more expensive.

As Amyris and São Martinho worked through the details of their December 2009 partnership, they determined that a joint venture structure would improve the likelihood of achieving near-term production. In April 2010, Amyris entered into a joint venture with Usina São Martinho, part of São Martinho SA, to build a new facility to turn out its products.[44] The joint venture, SMA Indústria Química SA, was created to build the first facility dedicated to the production of Amyris products, initially, farnesene, a molecule that can be used as an ingredient in a wide range of consumer and industrial items, including detergents, cosmetics, perfumes, and industrial lubricants. The farnesene produced by the mill will initially be sold to the consumer-products market because there it commands far higher prices than what is paid for diesel fuel. Under the agreement, Amyris Brasil will market and distribute the renewable; São Martinho will sell the feedstock and provide other services to the joint venture.

The construction costs for the new facility were estimated to total between $80 and $100 million. Under the terms of the joint venture agreement, construction of the facility will occur in two phases. Phase I is designed to build a facility capable of producing farnesene from up to one million tons of crushed sugarcane annually; Phase II will add up to two million tons of capacity. Using a joint venture arrangement enabled Amyris to reduce its capital costs. Within one year after the commencement of Phase I commercial operations, Usina São Martinho will reimburse Amyris for one half of the cost of the Phase I facility, up to an estimated cap of $17.5 million. Then, Usina São Martinho will co-fund Phase II and, as needed, make a final payment on the completion of construction so that its total contribution to the facility will equal an estimated $35 million.

In addition to scaling up its production capabilities in Brazil, Amyris also focused on patent protection, marketing, and securing additional funding. In November 2009, the firm obtained patent protection for its diesel and jet fuel gasoline, and chemical products.[45] The same month, it signed a memorandum of understanding with Embraer (Empreso Brasileira de Aeronáutica SA), General Electric, and Azul (Azul Linhas Aéreas Brasileiras) to evaluate the technical and sustainability aspects of the firm's renewable jet fuel made from sugarcane feedstock.[46] The initiative will culminate with a demonstration flight in 2012 of an Embraer E-Jet, using GE Aviation engines and belonging to Azul. At the same time, the Brazilian government through Financioadora de Estudos e Projectos contributed funding to Amyris Brasil's jet fuel project.

In December 2009, Amyris received a $24.3 million Integrated Biorefinery Program grant from the DoE to expand its pilot plant in Emeryville, California, to turn sweet sorghum into diesel fuel and petrochemicals.[47] Under the grant, Amyris must obtain an additional $10.5 million in expenses for the facility.

Building on its proven technology platform and its attempt to scale-up the commercialization of chemicals and biofuels, in September 2010 Amyris went public, raising $84.8 million. The firm will use the proceeds for capital expenditures related to establishing production facilities, including its joint venture facility in Brazil, and for general corporate purposes, including building engineering service capabilities to support Brazilian sugar mill adoption of its technology.

Amyris has embarked on an integrated strategy focused on flexible product lines. Its array of joint ventures, partnerships, and contract manufacturing agreements in the chemicals and fuels markets may enable the firm to pursue its "capital light" path.

135

LS9, Inc.

LS9, Inc. (LS9), a privately held industrial biotechnology company, pursues the commercial development of fuels and chemicals. The firm uses synthetic biology to make fuels and chemicals from the fermentation of cellulosic feedstocks, initially sugarcane.

Founded in 2005 by a Harvard Medical School geneticist—George Church, Ph.D., Director of the MIT–Harvard–DoE [Genomes to Life] Center—and a Stanford University plant biologist, Chris Somerville, Ph.D., it was officially launched in February 2007. Somerville thereafter left the firm, becoming head of the Energy Biosciences Institute, a partnership among the U.C. Berkeley, Lawrence Berkeley National Laboratory, and the University of Illinois, Champagne-Urbana. With Somerville's departure, Church became the firm's main scientific founder.

Scientific Background

LS9 is in the race to produce a biofuel product, specifically high-quality diesel fuel, and specialty chemicals.[48] The firm modifies the genetic makeup of microorganisms, including *E. coli*, a bacterium.

Similar to Amyris, LS9's business plan focuses on making its fuels in Brazil, where sugarcane is abundant and inexpensive and there exists good manufacturing capabilities, presently used for ethanol. Because feedstock accounts for about 80 percent of biofuels production expenses, using sugarcane keeps costs down.

The firm's one-step fermentation process uses patent-pending microbes to convert renewable plant-based materials, a feedstock, specifically sugarcane that can be broken down. The process uses the pathway that converts sugar into fatty acids that can then be diverted into diesel fuel. Similar to Amyris, but in contrast to others, such as SGI and Solazyme with multiple conversion steps, LS9's one-step conversion reduces production costs.

In January 2010, a research team, including LS9 scientists, members of the Jay Keasling laboratory at the U.C. Berkeley and the DoE's Joint BioEnergy Institute, announced a major breakthrough in their ability to make diesel and other biofuels directly from cellulosic biomass, plant matter composed of cellulose fibers, including crop residues, such as rice and wheat straw, corn stalks, and wood chips, in a one-step process.[49] Converting pretreated biomass in a one step to a renewable fuel, more technically, consolidated bioprocessing, represents a critical element in driving down the costs of cellulosic biofuels toward parity with conventional gasoline and diesel fuels. The single fermentation process does not require other carbon sources or additional chemical transformation, thereby lowering production costs. Technically, the researchers engineered the *E. coli* bacteria to express, in other words, produce hemicellulases, enzymes that break down, technically ferment, hemicellulose, the complex sugars that are a major component of cellulose biomass and the primary repository for the energy contained in plant cell walls, a key step toward helping bacteria produce biodiesel directly from cellulosic biomass. The team achieved a one-step process to create diesel fuel that requires no change in distribution or vehicle infrastructure to be deployed in a transportation fleet. The research was supported by funds from LS9 and the U.C. Discovery Grant program, the latter a three-way partnership among the University of California, private industry, and the State of California aimed at strengthening and expanding California's economy in targeted research fields.

Even before this research breakthrough, LS9 had already isolated and arranged the genes needed to make petroleum-like fuels in 2007 and introduced these genes into bacteria. Using synthetic biology tools, the company modified the genetic pathways in microbes that bacteria, plants, and animals naturally use to make fatty acids. These microbes break down sugarcane at a faster pace than normal. The sugarcane exudes fatty acids that consist of chains of carbon and hydrogen atoms strung together in a specific arrangement. Taking away the fatty acids from their carrier proteins unlocks the natural regulation that otherwise inhibits the making of additional fatty acids, resulting in an abundance of fatty acids.

The microbes start out as nonpathogenic strains of *E. coli* that LS9 modifies by redesigning their genome. Because crude oil is only a few molecular stages removed from fatty acids excreted by *E. coli* during fermentation, the desired result can now be achieved far more easily. As an added bonus, the firm does not kill its microbes to get the petroleum. The microbes secrete oil naturally and then live to feed, digest, and excrete more oil. Reusing the microbes instead of cultivating a new generation cuts the production time and decreases expenses.

Like Solazyme and Amyris, LS9's fuel product overcomes many operational and environmental challenges associated with conventional diesel and other biodiesel products. The fuel produced by LS9's microbes in the fermentation process is basically pump-ready, a "drop-in" fuel, needing only a simple cleaning step to filter out impurities via processing in a refinery. Its fuels are similar to the conventional forms so that they are compatible with existing fuel distribution infrastructure. They can be pumped through existing pipelines and directly into trucks and cars. The firm thereby bypasses one of biofuels, such as ethanol, biggest short-term problems, namely, distribution.

LS9 Ultra Clean fuels also offer environmental benefits in comparison to the production and refinement of crude oil or ethanol. Its fuels offer an 80–85 percent reduction in greenhouse gas emissions compared to conventional fuels and contain no carcinogens, such as benzene.[50] LS9's Ultra Clean Diesel, currently produced at the firm's pilot plant in South San Francisco, California, meets regulatory requirements for use in vehicles on roads in both the United States and Brazil. This fuel has received third-party validation and certification from the ASTM and from the Brazilian National Agency of Petroleum.[51]

Ramping-up Biofuels Production

To help validate the commercial viability of its technology and begin to commercialize its biodiesel product, in February 2010, LS9 announced the acquisition of the former Biomass Processing Technology production facility in Okeechobee, Florida, for $2 million.[52] LS9 will retrofit the facility to accommodate the firm's one-step fermentation process from nearby feedstocks, such as sugarcane. The facility will initially support the annual production of fifty thousand to one hundred thousand gallons of its Ultra Clean Diesel fuel and has the potential to produce ten million gallons per year of Ultra Clean Diesel. Furthermore, the facility will allow the firm to demonstrate that its one-step manufacturing process is ready and capable of bringing low-cost, low-carbon fuels to market and will house laboratory and pilot-scale operations for testing cellulosic materials, such as, wood chips and agricultural waste. Once retrofitted it will be the largest advanced biodiesel demonstration plant in the world.

Funding

LS9 has an impressive roster of venture capital funders. To date, it has raised $75 million in four rounds of venture capital financing.[53] The firm is named LS9 because it was the ninth life-science company launched by Flagship Ventures, one of its major venture capital providers.

In March 2007, during its first equity financing round, the firm raised $5 million from Flagship Ventures and Khosla Ventures, the latter firm is also a backer of Amyris. In October 2007, at the height of the stock market bubble, the company closed a second round of equity funding, obtaining $15 million from a group of institutional investors led by Lightspeed Venture Partners. Flagship Ventures and Khosla Ventures also participated in this round. The second funding round supported the firm's continued talent acquisition and the development of its biofuels products, including the construction of its California pilot facility. In September 2009, it completed a $25 million round of equity funding. Participating investors in this third round included CTV Investments LLC, the venture capital arm of Chevron Technology Ventures LLC, as well as previous investors, Flagship Ventures, Khosla Ventures, and Lightspeed Venture Partners. Then in December 2010, the firm raised an additional $30 million to ready its products for commercial production and support additional development and growth programs. All of its previous investors participated in this round.

Chemical Applications

LS9's fermentation process can also turn out chemicals used in making industrial and commercial items. To broaden its spectrum beyond fuels, in May 2009, the firm entered into a three-year strategic partnership with Procter & Gamble.[54] The partnership supports the joint development and commercialization of LS9's technology to produce chemicals for use in P&G's portfolio of consumer products, such as soaps and detergents. The multiyear arrangement will help LS9 accelerate its business plan and the adoption of its technology. Working with partners, such as P&G, having a broad commercial reach, will also assist LS9 to bring its technology to market faster. LS9's biodiesel will not, however, be commercially available until 2013, barring unforeseen difficulties or a lack of financing that may delay its arrival to market.

Problems Raised by Using Sugarcane as a Feedstock

Unlike algae-based firms, such as SGI, Sapphire, and Solazyme, Amyris and LS9 face several downsides in using sugarcane-based fuels. First, huge amounts of water are needed to grow the sugarcane feedstock. Although the requisite water comes from natural rainfall in Brazil, of greater concern is reducing the water intensity of the processing systems. Second, the environmental impact of growing sugarcane is uncertain. Some, such as, the ETC Group, express concern that biofuels produced from plant-based feedstocks, such as sugarcane, will cause environmental damage by eroding and degrading soils, reducing biodiversity, and increasing food insecurity.[55] In other words, these biofuels may be as environmentally destructive as the existing fossil-based petroleum system it replaces. Third, Brazilian farms have been steadily expanding and eating into the Amazon rainforest. It is unclear whether Amyris and LS9 can reach their production goals without converting any rainforest into farmland. However, sugarcane is only planted on about 3 percent of Brazil's viable land. It could easily expand onto the more than one hundred million acres presently used for grazing cattle in Brazil.[56]

Synthetic biology permits researchers to test numerous improvements in a short timeframe. Besides quick iterations, scientists do not stop innovating. Also, the economics of the next generation of biofuels and petrochemical technology will improve following scale-up, thereby enabling firms to take advantage of large-scale production efficiencies. The involvement of major venture capital firms, oil giants, and the U.S. military as well as the U.S. Departments of Energy and Agriculture will help sort out the various approaches. Major players with real expectations will separate approaches that will lead somewhere from those that will not. Any one company's technology may not perform as expected when applied at a commercial production level; it may face operational challenges for which it is unable to devise a workable solution. It may not be able to scale-up its production in a timely manner, if at all. It is uncertain how genetically modified algae or sugarcane will perform under commercial conditions. It is unlikely any one firm will beat out the others. With the demand for transportation fuels so huge, there likely will be multiple winners.

Notes

1.   Statement, Cynthia J. Warner, President, Sapphire Energy, Before the Senate Committee on Environment and Public Works, Hearing on Business Opportunities and Climate Policy, 111th Congress, 1st Session, May 19, 2009; Anne C. Mulkern, "Algae as Fuel of the Future Faces Great Expectations—and Obstacles," *Greenwire*, September 17, 2009, http://www.eenews.net/Greenwire/print/2009/09/17/1 (accessed April 19, 2010); Eilene Zimmerman, "Scum of the Earth," *Fortune Small Business* 19, no. 3 (April 2009): 60; Elizabeth Douglass, "From the Slime Emerges 'Green Crude,'" *Los Angeles Times*, May 29, 2008, C1; Kerry A. Dolan, "Turning Algae into Gasoline," *Forbes.com*, May 28, 2008, http://www.forbes.com/2008/05/28/alternative-fuels-biofuels-tech_sciences (accessed September 29, 2009); David Schwartz, "The A.I.M. Interview Sapphire Energy's

Tim Zenk," *Algae Industry Magazine.com*, February 9, 2010, http://www.algaeindustrymagazine.com/tim-zenk (accessed April 13, 2010).

2. Sapphire Energy, Inc. (Sapphire), Press Release, "Sapphire Energy Unveils World's First Renewable Gasoline," May 28, 2008. See also Douglass, "From the Slime"; Taylor Graham, "Sapphire Energy Pursues New Algae Biofuels Process to Make 'Green Crude,'" *Ethanol & Biodiesel News* 20, no. 23 (June 3, 2008), http://ProQuest (accessed November 9, 2009).

3. Continental Airlines, Press Release, "Continental Airlines Flight Demonstrates Use of Sustainable Biofuels as Energy Source for Jet Travel," January 7, 2009. See also Matthew L. Wall, "A Move toward Veggie Power Aloft," *New York Times*, January 7, 2009, B3; Bill Hensel Jr. "Biofuels Take Off," *Houston Chronicle*, January 8, 2009, Business Section, 1; Dan Reed, "Which of Those could Fuel a Jet? A B C," *USA Today*, January 27, 2009, 7A; David Biello, "For Greening Aviation, Are Biofuels the Right Stuff?" *Yale Environment* 360, June 11, 2009, http://e360.yale.edu/content/feature.msp?id=2160 (accessed November 12, 2009).

4. Sapphire, Investors, http://www.sapphireenergy.com/investors (accessed November 9, 2009); Sapphire, Securities and Exchange Commission (SEC) Form D, June 20, 2007 and December 21, 2007; Sapphire, Press Release, "Sapphire Energy Builds Investment Syndicate to Fund Commercialization of Green Crude Production," September 17, 2008. See also Chris Morrison, "Sapphire Energy Gets 'Open Checkbook' from Investors for Algae-Based Gasoline," *Green Beat*, http://green.venturebeat.com/2008/05/29 (accessed September 29, 2009); Matt Levin, "Algae Power," *Private Equity International* 67 (July–August 2008): 40; Martin LaMonica, "Bill Gates Invests in Algae Fuel," *Green Tech*, September 17, 2008, http://news.cnet.com/8301-11128_3-10043996-54.html (accessed March 23, 2010). In March 2011, Monsanto made an equity investment in Sapphire and the two firms will use Sapphire's technology platform to discover genes in algae to improve crop yields and stress tolerance. Monsanto Co., Press Release, "Monsanto Company and Sapphire Energy Enter Collaboration to Advance Yield and Stress Research," March 8, 2011.

5. U.S. Department of Energy, Press Release, "Secretaries Chu and Vilsack Announce More Than $600 million Investment in Advanced Biorefinery Projects," December 4, 2009; U.S. Representative Harry Teague, Press Release, "Recovery Act Project Will Create Clean Energy Jobs in Southern New Mexico," December 4, 2009. See also Michael Hartranft, "Gold in Green," *Albuquerque Journal* (New Mexico), December 5, 2009, D3; Kevin Robinson-Avila, "Feds Fund Sapphire Energy Algae Project," *New Mexico Business Weekly*, December 4, 2009, http://albuquerque.bizjournals.com/albuquerque/stories/2009/11/30/daily50.html (accessed March 23, 2010).

6. Sapphire, Press Release, "Algae-Based Fuel Projected to be Commercial-Ready in Three Years," April 16, 2009. See also Geoffrey Thomas, "Biofuel Paradox," *Air Transport World* 47, no. 1 (January 1, 2010): 47–49, at 48; Drake Lundell, "Algae-to-Fuel Forum Precedes World Biofuels Market 2009 Program," *Ethanol & Biodiesel News* 21, no. 11 (March 17, 2009), http://ProQuest (accessed November 9, 2009).

7. I have drawn on Solazyme, Inc. (Solazyme), *Technology*, http://www.
    solazyme.com/technology (accessed September 24, 2009); Solazyme, *Fuels
    + Chemicals*, http://www.solazyme.com/market/fuels (accessed September
    24, 2009); Solazyme, *Health Sciences*, http://www.solazyme.com/mar-
    ket/health (accessed September 24, 2009); Solazyme, *Human Nutrition
    and Animal Nutrition*, http://www.solazyme.com/market/nutritionals
    (accessed September 24, 2009). Solazyme, Preliminary SEC Form S-1,
    March 11, 2011, 2, 44–5, 68–9; Harrison Dillon, "Striking Algal Oils,"
    Chemistry and Industry 19 (October 11, 2010): 18–20. See also Kevin Bul-
    lis, "Fuel from Algae," *Technology Review*, February 22, 2008, http://www.
    technologyreview.com/business/20319/?a=f (accessed April 21, 2010);
    Lindsay Riddell, "Algae can Make the World Go Round," *San Francisco
    Business Times*, June 12–18, 2009, 15; Geoffrey Brooks, "Sustainable Mi-
    croalgae Herald the Future," *Functional Ingredients* 82 (December 2008):
    30–33; Melody Voith, "Up from the Slime," *Chemical & Engineering News*
    87, no. 4 (January 26, 2009): 22–23; Liz Turner, "After 30 Years, Algae-
    to-Fuel Finally Gets the Green Light," *Green Fuels Forecast*, March 2008,
    http://www.greenfuels.forecast.com/ArticleDetails.php?articleID=481
    (accessed November 9, 2009).

    In November 2010, Solazyme and Roquette Frères SA entered into a
    joint venture agreement for the production, commercialization, and market
    development of microalgae-derived food ingredients and nutraceuticals.
    Solazyme, Press Release, "Solazyme and Roquette Sign Agreement to Cre-
    ate Global Nutritional Joint Venture," November 8, 2010 and Solazyme,
    Preliminary SEC Form S-1, 47, 80, 92–3, F-32. In March 2011, the firm
    announced an agreement with Dow Chemical to jointly develop Solazyme's
    algal oils for dielectric insulating fluids and a nonbinding letter of intent
    giving Dow the option of buying specified quantities of the company's oils.
    Solazyme, Press Release, "Solazyme and Dow Form an Alliance for the
    Development of Micro Algae-Derived Oils for Use in Bio-based Dielectric
    Insulating Fluids," March 9, 2011 and Solazyme, Preliminary SEC Form
    S-1, 47, 78, 95.

8. Solazyme, Press Release, "Solazyme Produces First Algal-Based Renewable
    Diesel to Pass American Society for Testing and Materials D-975 Specifica-
    tions," June 11, 2008.

9. Solazyme, Press Release, "Solazyme Showcases World's First Algal-Based
    Renewable Diesel at Governor's Global Climate Summit," November 17,
    2008; Solazyme, Press Release, "Solazyme Unveils Renewable Biodiesel
    Derived from Algae via Scalable Process," January 22, 2008.

10. "Beyond Fossil Fuels: Harrison Dillon on Biofuels," *Scientific American*,
    April 22, 2008, http://www.scientificamerican.com/article.cfm?id=energy-
    dillon-solazyme (accessed September 3, 2009); Solazyme, Press Release,
    "Additional Emission Testing Demonstrates Solazyme's Algal-Biofuels
    Shows to Significantly Lower Tailpipe Emission When Compared to Ultra-
    Low Sulfur Diesel," April 21, 2009.

11. Solazyme, Press Release, "Solazyme Produces World's First Algal-Based
    Jet Fuel—Fuel Passes All Tested Specifications including the Most Criti-
    cal ASTM DI655 Specifications," September 9, 2008. See also Kathleen

Koster, "Solazyme Microbial Diesel Passes ASTM Jet Fuel Test," *Ethanol & Biodiesel News* 20, no. 37 (September 9, 2008), http://ProQuest (accessed November 9, 2010). In February 2011, Solazyme announced a collaboration with Qantas to pursue the commercial production of Solajet in Australia. Solazyme, Press Release, "Solazyme and Qantas Launch Collaboration Working toward Commercial Production of Solajet," February 10, 2011.

12. Solazyme, Press Release, "Solazyme Wins Navy Contract to Provide World's First 100% Algal Based Jet Fuel," September 24, 2009. By July 2010, Solazyme completed delivery of 1,500 gallons of algal jet fuel to the U.S. Navy for testing. Solazyme, Press Release, "Solazyme Delivers 100% Algal-Derived Renewable Jet Fuel to U.S. Navy," July 19, 2010. In September 2010, Solazyme completed delivery of more than twenty thousand gallons of algal-derived shipboard fuel to the U.S. Navy. Solazyme, Press Release, "Solazyme Completes World's Largest Microbial Advanced Biofuel Delivery to U.S. Military," September 15, 2010. Then in October 2010, the U.S. Navy successfully tested a vessel powered by a 50-50 blend of the Navy's traditional shipboard fuel and Solazyme's algae-based renewable diesel fuel. Solazyme, Press Release, "Navy Demonstrates Solazyme's Soladiesel Renewable F-76 Fuel," October 22, 2010.

13. Solazyme, Press Release, "Solazyme Signs U.S. Department of Defense Contract to Develop Navy Fuels from Algae," September 8, 2009; Defense Energy Support Center, Press Release, "New Energy Solution Emerging for Warfighters," September 3, 2008. See also Lindsay Riddell, "Solazyme Wins $8.5m Navy Contract," *San Francisco Business Times*, September 8, 2009, http://sanfrancisco.bizjournals.com/sanfrancisco/stories/2009/09/07/daily14.html (accessed November 9, 2009).

14. Harrison Dillon in "Algae Production Systems: Opportunities & Challenges," *Industrial Biotechnology* 5, no. 4 (December 22, 2009): 216–26, at 222 and Solazyme, Preliminary SEC Form S-1, 1, 44, 72, 82–4.

15. Solazyme, Press Release, "Solazyme Testing Blue Fire Ethanol Cellulosic Sugars in Its Microalgae Renewable Oil Production Process," May 26, 2009. See also Michael Kanellos, "The Anticipated Market for Feedstock Sugars Takes Root," *Greentech Media*, May 26, 2009, http://www.greentechmedia.com/green-light/post/the-anticipated-market-for-feedstock (accessed November 13, 2009).

16. Solazyme, *Technology*; Dillon in "Algae Production Systems," 221. See also "Beyond Fossil Fuels," *Scientific American* ; Bullis, "Fuel from Algae."

17. Solazyme, Press Release, "Solazyme Completes Financial Round," December 21, 2005.

18. Solazyme, SEC Form D, February 23, 2007 and May 11, 2007; Harris & Harris Group, Inc., Press Release, "Harris & Harris Group Invests in Solazyme," March 7, 2007; Harris & Harris Group, Inc., 2008 Annual Report, 45, 47, 70, 82.

19. BlueCrest Capital Finance, L.P, Press Release, "BlueCrest Capital Finance Supplies $5 Million in Debt Financing to Solazyme, Inc.," September 19, 2007 and Solazyme, Preliminary SEC Form S-1, F-24.

20. Solazyme, Press Release, "Latest Funding Brings Total Series C Round to $57 Million," June 5, 2009; Solazyme, SEC Form D, March 11, 2009, September 29, 2008, and August 8, 2008; Solazyme, Press Release, "Solazyme

Announces Series D Financing Round of More Than $50 Million," August 9, 2010; Solazyme, Press Release, "Solazyme Adds Bunge as Strategic Investor in Series D," August 26, 2010; Solazyme, Press Release, "Solazyme Adds Unilever as Strategic Investor," September 7, 2010; Solazyme, Press Release, "Solazyme Adds Sir Richard Branson as Strategic Investor," September 8, 2010. For a summary of the firm's issuance of redeemable convertible preferred stock, see Solazyme, Preliminary SEC Form S-1, F-20 to F-24. See also Peter Ngo, "Solazyme Raises $45M in Private Placement," *Ethanol & Biodiesel News* 20, no. 36 (September 2, 2008), http://ProQuest (accessed November 9, 2009); Emma Rich, "Solazyme Joins Algae Elite with Additional $45m," *CleanTech*, August 26, 2008, http://cleantech. com/news/print/3306 (accessed November 12, 2009); Paul Sonne, "To Wash Hands of Palm Oil Unilever Embraces Algae," *Wall Street Journal*, September 8, 2010, B1; Yuliya Chernova, "Going Commercial with Algae," *Wall Street Journal*, September 12, 2010, B6.

21. Solazyme, Press Release, "Solazyme Selected for National Institute of Standards and Technology Award," September 27, 2007 and Solazyme, Preliminary SEC Form S-1, F-24.

22. Solazyme, Press Release, "Solazyme and Chevron Technology Ventures Enter into Biodiesel Feedstock Development and Testing Agreement," January 22, 2008 and Solazyme, Preliminary SEC Form S-1, 47, 93–4, F-22 to F-23. See also David R. Baker, "Startup Partners with Chevron on Algae Fuel," *San Francisco Chronicle*, January 23, 2008, C1.

23. U.S. Department of Energy, Press Release, "Secretaries Chu and Vilsack Announce" and Solazyme, Preliminary SEC Form S-1, 85, 91, F-23. See also "Solazyme Leading Renewable Oil Company Awarded $21 million Energy Grant for Biorefinery Project in Pennsylvania," *Biotech Business Week*, December 28, 2009 (accessed February 1, 2010). Joseph N. DiStefiano, "Solazyme Gets Grant for Biorefinery Project in Pa.," *Philadelphia Inquirer*, December 8, 2009, E1.

24. Solazyme, Press Release, "Solazyme and Unilever Partner to Bring Algal Renewable Oil to Personal Care Products," March 9, 2010 and Solazyme, Preliminary SEC Form S-1, F-23. See also "Here I am to Save the Day: Microalgae Gains Momentum with a Mighty March," *Biofuels Digest*, April 6, 2010, http://www.biofuelsdigest.com (accessed April 7, 2010). In March 2011, Solazyme announced agreements with Sephora and QVC to launch a microalgae-based anti-aging skincare line. Solazyme, Press Release, "Solazyme, Sephora and QVC Announce Agreements for the U.S. and International Launch of Solazyme's New Anti-Aging Skincare Line, Algenist," March 7, 2011 and Solazyme, Preliminary SEC Form S-1, 47, 82, 94–5, F-24.

25. For background on Keasling, see Alice Damil, "HERO by Nature," *Prism* 16, no. 9 (Summer 2007): 34–37 and Tom Abate, "Biofuel guru," *San Francisco Chronicle*, November 21, 2010, A1.

26. Hiroko Tsuruta, "High-Level Production of Amorpha-4, 11-Diene, a Precursor of the Antimalarial Agent Artemisinin, in *Escherichia coli*," *PLoS One* 4, no. 2 (February 2009): e 4489, chronicled steps taken to achieve the production of amorphadiene, a precursor of artemisinin, through

E. coli fermentations. The interplay of industrial fermentation processes and synthetic biology achieved the required increase in amorphadiene production levels. See also Victoria Hale et al., "Microbially Derived Artemisinin: A Biotechnology Solution to the Global Problem of Access to Affordable Anitmalarial Drugs," *American Journal of Tropical Medicine and Hygiene* 77, Suppl. 6 (2007): 198–202, who described the technological platform to manufacture semisynthetic artemisinin through fermentation and Jennifer R. Anthony et al., "Optimization of the Mevalonate-Based Isoprenoid Biosynthetic Pathway in *Escherichia coli* for Production of the Anti-Malarial Drug Precursor Amorpha-4, 11-diene," *Metabolic Engineering* 11, no. 1 (January 2009): 13–19.

27. Dae-Kyun Ro et al., "Production of the Antimalarial Drug Precursor Artemisinic Acid in Engineered Yeast," *Nature* 440, no. 7086 (April 13, 2006): 940–43. See also Amyris, Press Release, "U.C. Berkeley and Amyris Biotechnologies Collaborate to Demonstrate Microbial Production of Artemisinin," n.d.

28. Amyris Biotechnologies, Inc. (Amyris), Press Release, "$42.6 million Five-Year Grant from Gates Foundation for Antimalarial Drug Brings Together Unique Collaboration," n.d.; Institute for OneWorld Health, Press Release, "$42.6 million Five-Year Grant Brings Together Unique Collaboration of Biotech, Academia, and Nonprofit Pharma," December 14, 2004. See also Paul Jacobs, "Grant to Aid San Francisco Nonprofit's Production of Malaria Treatment," *San Jose Mercury News*, December 13, 2004, IE; Vivien Marx, "Bootstrapping via Philanthropy," *Chemical & Engineering News* 83, no. 1 (January 3, 2005): 18; Renuka Rayasam, "Bets on Biotech," *U.S. News & World Report* 141, no. 11 (September 25, 2006): 44–45. See also Tracy Hampton, "Collaboration Hopes Microbe Factories Can Supply Key Antimalaria Drug," *Journal of American Medical Association* 293, no. 7 (February 16, 2005): 785–87.

29. Steve Johnson, "Low-Cost Malaria Drug in Sight," *San Jose Mercury News*, April 13, 2006, Business Section, 1.

30. Amyris, Press Release, "OneWorld Health, Amyris and Sanofi-aventis Announce Development Agreement for Semisynthetic Artemisinin," March 3, 2008.

31. Jason Pontin, "First, Cure Malaria. Next, Global Warming," *New York Times*, June 3, 2007, Business Section, 3. See generally, David R. Baker, "Amyris Brews Barrels of Diesel from Sugarcane," *SFGate*, November 12, 2008, http://www.sfgate.com/cgi-bin/article.cgi?f=/c/a/2008/11/12 (accessed September 29, 2009); Emma Rich, "Microbes Drive New Amyris Biodiesel Plant," *CleanTech*, November 11, 2008, http://www.cleantech.com/news/3858/microbes-drive-new-amyris-biodiesel-plant (accessed September 29, 2009).

32. Amyris, Preliminary SEC Form S-1, April 16, 2010, 11, 44, 75. In June 2010, Amyris entered into a series of partnerships and agreements, including a partnership with Soliance for the production and commercialization of renewable cosmetic ingredients, a joint venture agreement with Cosan SA for the production and commercialization of renewable intermediate chemicals, a two-part collaboration agreement with M&G Finanziaria SRL,

a series of agreements with Procter & Gamble Co., and an agreement with Shell Western Supply and Trading Ltd., a subsidiary of Royal Dutch Shell PLC. Amyris, Press Release, "Amyris and Soliance Partner to Commercialize Renewable Bio-Sourced Cosmetics," June 22, 2010; Amyris, Press Release, "Amyris and Cosan Creating JV for Production and Commercialization of Cane Based Renewable Chemicals," June 22, 2010; Amyris, Press Release, "Amyris and M&G Finanziaria Enter into Off-Take Agreement," June 24, 2010; Amyris, Press Release, "Amyris Enters into Multi-Products Collaboration and Off-Take Agreement with the Procter and Gamble Company," June 24, 2010; Amyris, Press Release, "Amyris Enters into Off-Take Agreement with Shell," June 25, 2010; Amyris, Amendment No. 2 to SEC Form S-1, June 22, 2010, 2, 15, 51, 61, 76, 81, 87, 88, 90.

After its initial public offering, the firm announced a series of manufacturing agreements with: Tate & Lyle Ingredients, Inc., an affiliate of Tate & Lyle PLC, to produce farnesene at its facilities in Decatur, Illinois; Antibióticos SA to produce farnesene at its facilities in León, Spain; and Paraíso Bioenergia SA pursuant to which Amyris will construct fermentation and separation capacity in Brazil to produce its products. Amyris, Press Release, "Amyris Signs Contract Manufacturing Agreement with Tate & Lyle," November 4, 2010; Amyris, Press Release, "Amyris Contracts for Biofene Manufacture with Antibióticos, S.A.," March 3, 2011; Amyris, Press Release, "Amyris Adds Production Capacity through Agreement with Paraíso Bioenergia," March 22, 2011.

33. Amyris, Press Release, "Amyris Biotechnologies Raises $20 million in Series of Funding," n.d., Amyris, SEC Form D, May 25, 2007; Amyris, Press Release, "Amyris Announces $70 Million Series B Round," September 19, 2007; Amyris, SEC Form D, September 18, 2007. For a summary of Amyris private placements, see Amyris, Preliminary SEC Form S-1, 110.

34. Amyris, Press Release, "Amyris Secures Funding to Advance Scale Up, First Commercial Plant," October 1, 2009; Amyris, SEC Form D, August 18, 2009 and August 20, 2009. See also Lisa Sibley, "Friendly Competitor LS9 Congratulates Amyris on $25M Raise," *CleanTech*, August 21, 2009, http://www.cleantech.com/news/print/4878 (accessed March 15, 2010).

35. Amyris, Press Release, "Amyris Names Scientist and Entrepreneur Fernando Reinach to Board of Directors," October 14, 2008.

36. Amyris, "Amyris Biotechnologies Secure Investment from Temasek Holdings," March 31, 2010 and Amyris, "Total and Amyris to Build a Strategic Partnership for Biomass-Based Fuels and Chemicals," June 23, 2010.

37. Amyris, Press Release, "Amyris and Crystalsev Join to Launch Innovative Renewable Diesel from Sugarcane by 2010," April 28, 2008. See also Matt Nauman, "Biotech Deal for Clean Fuel," *San Jose Mercury News*, April 24, 2008, 1C; Kris Bevill, "Amyris, Crystalsev form Joint Venture to Produce Renewable Diesel by 2010," *Biodiesel Magazine*, May 2008, http://www.biodieselmagazine.com/article-print (accessed March 15, 2010); David Ehrlich, "Amyris, Crystalsev in Sugarcane Biodiesel Venture," *CleanTech*, April 24, 2008, http://www.cleantech.com/news/print/2757 (accessed March 15, 2010).

38. Amyris, Press Release, "Amyris Opens Pilot Plant to Produce Renewable Diesel Fuel," November 12, 2008. See also Emma Ritch, "Microbes Drive

New Amyris Biodiesel Plant," *CleanTech*, November 11, 2008, http://www.cleantech.com/news/print/3858 (accessed March 17, 2010).

39. Amyris, Press Release, "Amyris Renewable Diesel Receives EPA Registration," April 20, 2009; Amyris, Preliminary SEC Form S-1, 81; Amyris, Press Release, "Amyris No Compromise Renewable Diesel Receives Highest EPA Blending Registration," November 1, 2010.

40. Amyris, Press Release, "Amyris Opens Renewable Products Demonstration Facility in Brazil," June 25, 2009; Amyris, Preliminary SEC Form S-1, 2.

41. Amyris, Press Release, "Amyris and São Martinho Group Enter Into Agreement," December 3, 2009. See also "Amyris to Acquire Stake in Brazilian Ethanol Mill," *Industrial Biotechnology* 5, no. 4 (December 22, 2009): 195.

42. Mara Lemos Stein, "Stratus Cleantech Leads $80M Round in Amyris Brasil," *LBO Wire*, March 4, 2010, http://ProQuest (accessed April 22, 2010).

43. Amyris, Press Release, "Amyris Signs Letters of Intent Agreements with Bunge, Cosan and Guarani," December 8, 2009; Amyris, Preliminary SEC Form S-1, 2, 45, 68. See also "Amyris Brasil Developing Molasses-Based Diesel," *Industrial Biotechnology* 5, no. 4 (December 22, 2009): 195; "Amyris Signs Letters of Intent Agreements with Bunge, Cosan and Guarani," *Biotech Business Week* (December 28, 2009): 719, http://LexisNexis (accessed February 1, 2010).

44. Amyris Press Release, "Amyris and São Martino Group Establish Joint Venture at Usina São Martinho," April 15, 2010; Amyris, Preliminary SEC From S-1, 13, 46. In April 2011, Amyris announced the completion of its first industrial-scale production facility in Brazil. Amyris, Press Release, "Amyris's First Commercial Production Facility Complete and Operational," April 29, 2011.

45. Amyris, Press Release, "Amyris Obtains Patents for No Compromise® Renewable Fuels and Chemical Products," November 17, 2009; Amyris, Preliminary SEC Form S-1, 82. See also "Amyris Announces Technology-Portfolio Patents," *Industrial Biotechnology* 5, no. 4 (December 22, 2009): 195.

46. Amyris, Press Release, "Embraer, General Electric, Azul, and Amyris Announce Renewable Jet Fuel Evaluation Project," November 18, 2009. See also Megan Kuhn, "Azul Looks to Sugar Cane to Power Embraer E-Jet," *Flight International* 176, no. 5216 (November 24–30, 2009): 12.

47. U.S. Department of Energy, "Secretaries Chu and Vilsack Announce." See also Andrew S. Ross, "No Cheers for Wine Industry," *San Francisco Chronicle*, December 8, 2009, D1.

48. LS9, Inc. (LS9), Renewable Petroleum™ Technology, http://www.ls9.com/technology (accessed February 25, 2010); LS9, UltraClean™ Fuels, http://www.ls9.com/products/chemical.html (accessed February 25, 2010). See also Elizabeth Svoboda "Fueling the Future," *Fast Company* 122 (February 2008): 45–47.

49. LS9, Press Release, "LS9, Inc., U.C. Berkeley, and JBEI Make Major Breakthrough in Cellulosic Fuels Production," January 27, 2010. See also Andrew Turley, "Synthetic Biology Breakthrough may Fuel Future," *Chemistry and Industry* 3 (February 8, 2010): 5–6; Eric J. Steen et al., "Microbial Production

of Fatty-Acid-Derived Fuels and Chemicals from Plant Biomass," *Nature* 463, no. 7280 (January 28, 2010): 559–62; "JBEI, LS9 Reengineer *E. coli* to Produce Renewable Diesel Directly from Biomass," *Biofuels Digest*, January 28, 2010, http://www.biofuelsdigest.com/blog2/2010/01/28/jbei-ls9-reenginer-e-coli-to-produce-ren (accessed March 11, 2010).

50. LS9, UltraClean™ Fuels.

51. LS9, Press Release, "LS9 UltraClean™ Diesel Exceeds United States & Brazilian Fuels Specifications," July 20, 2009. See also Lia Sibley, "LS9's Clean Diesel Meets U.S., Brazilian Requirements," *CleanTech*, July 20, 2009, http://www.cleantech.com/news/print/4724 (accessed September 13, 2009); Svoboda, "Fueling the Future."

52. LS9, Press Release, "LS9 Purchases Florida Site to Manufacture Renewable Petroleum," February 3, 2010. See also Lisa Sibley, "LS9 Scores Manufacturing Site for a Bargain $2M," *CleanTech*, February 2, 2010, http://www.cleantech.com/news/print/5593 (accessed March 10, 2010); Camille Ricketts, "Biofuel Leader LS9 Buys Demo Plant to Churn Out Renewable Diesel," *Venture Beat*, February 3, 2010, http://www.venturebeat.com/2010/02/03/biofuels-leader-ls9-buys-demo-plant-to-churn-out-renew (accessed March 10, 2010); Mara Lemos Stein, "Renewable-Fuel Company LS9 Buys Plant from BP Technology," *Daily Bankruptcy Review Small-Cap*, February 4, 2010, http://ProQuest (accessed February 12, 2010).

53. LS9, Press Release, "Renewable Petroleum Company LS9 Announces Series A Funding and Scientific Advisory Board Members"; LS9, SEC Form D, October 7, 2007; LS9, Press Release, "LS9, Inc. Secures $15 million in Series B Funding," October 9, 2007; LS9, Press Release, "LS9 Secures $25 Million in Latest Round of Funding," September 24, 2009. LS9, Press Release, "LS9 Raises $30 Million to Ready Products for Commercial Production," December 21, 2010. See also Michael Kanellos, "Chevron Invests in LS9," *GreenTechMedia*, September 24, 2009, http://ProQuest (accessed March 10, 2010).

54. LS9, Press Release, "LS9 and Procter & Gamble Launch Sustainable Chemical Partnership," May 19, 2009. See also Andrew S. Ross, "P & G Switching to 'Designer Microbes' to Replace Petrochemicals in Products," *San Francisco Chronicle*, May 21, 2009, C1; Lisa Sibley, "LS9, P & G Team up for Sustainable Chemicals," *CleanTech*, May 19, 2009, http://www.cleantech.com/news/print/4474 (accessed September 13, 2009); Lindsay Riddell, "LS9 Inc. Links with Procter & Gamble," *San Francisco Business Times*, May 19, 2009, http://www.sanfrancisco.bizjournals.com/sanfrancisco/stories/2009/05/18 (accessed March 10, 2010). In February2011, LS9 announced a second development and commercialization partnership with P&G. LS9, Press Release, "LS9 Announces Second Partnership with Procter & Gamble," February 7, 2011.

55. ETC Group, "Who Owns Nature? Corporate Power and the Final Frontier in the Commodification of Life," November 2008, 39; ETC Group, "Commodifying Nature's Last Straw? Extreme Genetic Engineering and the Post-Petroleum Economy," October 2008.

56. Antonio Regaldo, "Searching for Biofuels' Sweet Spot," *Technology Review* 113, no. 2 (March/April 2010): 46–51, at 50.

9

Conclusion: Policy Implications of Synthetic Biology Research and Commercialization

Synthetic biology research and commercialization raises three major policy issues: safety; bioterrorism; and the creation of "unfair" monopolies based on intellectual property rights. In approaching each of these areas, policymakers need to devise solutions to minimize the risks from harmful uses while minimizing the impediments to the beneficial uses. The risks of runaway microbes that could wreak havoc, for instance, must be balanced against the potential benefits flowing from the development and application of synthetic biotechnologies including new sources of fuel, chemicals, therapeutics, and environmental remediation. In striving to minimize the possible harms while facilitating scientific and societal progress, at present, I see no need to treat synthetic microorganisms as dangerous until proven harmless[1] or to impose a moratorium on experimentation, release, and commercialization.[2] Given the momentum, the wide dissemination, and the global character of the research, it is too late to impose these restrictive measures; however, finding the "right" balance between governmental regulation and self-regulation by the science community remains a difficult, often contentious, issue.

Safety Aspects

Concerns exist about potential risks to public health and the environment from accidental releases of genetically engineered microbes and from intentional nonlaboratory contained uses. Faced with uncertain and inadequate information regarding complex synthetic organisms, policymakers and regulators may err on the side of caution

149

and refuse to approve a process or a product thereby foregoing the societal benefits. Or, policymakers and regulators may err on the side of innovation and economic benefit, risking possible harmful consequences to public health and the environment. In any event, even if they possess the requisite legal authority, regulators must develop the requisite expertise and capabilities in a cutting-edge scientific field.

Today, using DNA synthesizers, the four subunit bases can be assembled to form DNA in any sequence from readily available chemicals and related substances used in laboratory processes. A genome-length stretch of DNA can be assembled from smaller pieces of DNA, called oligonucleotides or oligos. Oligos can be ordered from a commercial oligonucleotide manufacturer or made in a laboratory using a specialized machine.

Regulation of Recombinant DNA Research

The current safety concerns regarding synthetic biology harken back to the mid-1970s when biologists first discovered how to transfer a gene between organisms. A group of leading scientists raised concerns about the safety of rDNA research and called for a moratorium on certain kinds of experimentation in what is known as the "moratorium letter" until the development of safety guidelines and more experience could be gained in risk assessment.[3]

The February 1973 meeting of 140 scientists on rDNA, at the Asilomar Conference Center in Pacific Grove, California, established the foundation for biosafety based on the containment of microbes used in research. Excluding broader social and ethical concerns, the major biosafety issue discussed at Asilomar focused on the safety of transmitting genes from one organism to another organism via a third organism, such as a virus or bacterium. At the conference, the scientists called for the development of safety guidelines and a process for reviewing proposed rDNA experiments,[4] matching the containment strategy to the level of risk posed by the material worked on, based on a classification of risk levels. They also put forward ideas about a national policy group as an advisory body to the NIH. Today, nearly four decades later, we know that rDNA research can be performed safely. Many of the restrictions put in place after the Asilomar conference were unnecessarily restrictive because the potential biosafety risks of microbes engineered through rDNA technology were overstated.[5]

The Asilomar recommendations led, however, to two significant developments: research guidelines and an oversight mechanism.

First, the recommendations provided the basis for the 1976 National Institutes of Health Guidelines for Recombinant DNA Research (NIH Guidelines) and their subsequent revisions.[6] In brief, the guidelines seek to assess the risk of proposed research and then define an appropriate level of physical and biological containment. The containment measures must be proportionate to an experiment's risk characteristics, with increasing containment levels required for research that is more pathogenic to humans. Today, the NIH Guidelines form the principal line of defense against the accidental release of harmful genetically engineered organisms from laboratories. The long and generally safe record of laboratories in handling potentially dangerous materials provides assurance that researchers possess the capacity to protect workers and the surrounding community from the accidental release of deleterious, genetically engineered microorganisms.

Although the NIH Guidelines apply only to NIH-funded research, other federal science funding agencies, such as the Department of Defense and the Department of Agriculture, have incorporated the guidelines by reference into their own grants. The NIH Guidelines do not apply to privately funded research; however, the guidelines provide a means for voluntary industry compliance.[7] In addition, private laboratories must comply with general federal and state public health and environmental laws. Under common (case) law, these firms are liable to compensate employees and the public for damages caused by negligent activities, including the improper handling of potentially hazardous substances. In negligence lawsuits, the NIH Guidelines, among other sources, provide a reasonable standard of care for researchers. Thus, the guidelines constitute a general standard of practice and care for biosafety risk assessment and management in the area of rDNA research and commercialization.

Second, the Asilomar recommendations led to the formation of the NIH Recombinant DNA Advisory Committee (RAC) to oversee the safety of rDNA research and to establish appropriate standards for containing potentially risky experiments. The RAC was designed to provide independent federal scientific oversight of rDNA research. Although the RAC was conceived as a scientific review process, congressional and public concerns led to the appointment of nonscientific representatives to the RAC. As experience with genetic technologies grew, the NIH Guidelines were relaxed and experiments were conducted with minimal risk. Gradually, the RAC delegated much of its review authority for rDNA experimentation proposals to local institutional biosafety committees (IBCs). Today, IBCs review most rDNA research at NIH-funded institutions with only certain experiments requiring review and approval by the RAC or the NIH Director.

Under NIH rules, each research institution receiving NIH funding must institute biosafety procedures and establish an IBC. Each IBC reviews rDNA research conducted at the institution, approves certain research proposals, and ensures that they comply with the NIH Guidelines. Each IBC must consist of not less than five people, including at least two members not affiliated with the institution, and collectively the group must contain the appropriate rDNA expertise. If required by the nature of the research, an IBC may consult with ad hoc experts.

Depending on the organism used and the risk level of the proposed experiment, the principal investigator may be required to notify the IBC or obtain IBC approval before commencing the research. In reviewing or approving the proposed rDNA research, an IBC determines the appropriate physical and biological containment levels, using the NIH Guidelines, thereby ensuring adequate biosafety safeguards. In certain cases involving novel issues or higher risk levels, prior review and approval by RAC or the NIH Director may be required. NIH-funded institutions must register their respective IBCs with the NIH's Office of Biotechnology Activities, provide a roster of IBC members, their backgrounds, and update the NIH annually. IBCs must meet regularly, keep minutes, and open their meetings to the public, when consistent with the protection of privacy and proprietary interests. Researchers who choose to do so can bypass both the NIH Guidelines and the IBC review process by obtaining nonfederal funding.

In 1986, as the first commercial rDNA products emerged from the laboratories into field testing, the White House Office of Science and Technology Policy issued a Coordinated Framework for the Regulation of Biotechnology (Coordinated Framework) in the United States.[8] The Coordinated Framework, reflecting the then-scientific consensus, stated that rDNA technology did not present any unique risks or pose any problems that differed from those of conventionally produced organisms. Genetic engineering involved the insertion (or removal) of a small number of genes, leaving the host organism largely intact. Thus, the focus of federal governmental regulation, according to the Coordinated Framework, ought to be on the risk characteristics of the final product, not the production process. Furthermore, the existing U.S. regulatory framework could deal with any potential risks associated with biotechnology products likely to be developed, at least in the foreseeable future.

As a result, the current federal government review process for biotechnology products uses the same laws and regulations that apply to conventional goods, with the type of regulatory review dependent on the specific item. In brief, three federal agencies, the FDA, the EPA, and the USDA, have assumed the primary responsibility for regulating genetically engineered organisms and products under some dozen different laws. The laws under which these three governmental components rely for their regulatory authority over biotechnology products are more general laws, enacted for other purposes.[9] Over the years, the FDA, EPA, and USDA have issued guidelines and regulations as needed to clarify the application of existing laws to biotechnology products.

Regulation of Synthetic Biology Safety

With synthetic biology, researchers can engineer more complex genetic modifications than can be achieved through standard genetic engineering techniques. Policymakers and regulators need to figure out how to assess the safety of organisms with extensively modified genomes, perhaps derived from a large number of initial sources, even, hundreds. It may be difficult to determine an organism's genetic pedigree if it is assembled from multiple sources or if it contains DNA that is extensively manipulated, not merely transferred from somewhere else in nature. Synthetic biology may eventually enable the creation of artificial organisms with genetic elements designed from scratch. Even today, however, the various genetic elements may be very different from those created through existing genetic engineering techniques, thereby raising three difficult questions. First, will the new organisms created by synthetic biology present new or enhanced risks compared to those of today's genetic engineering techniques? Second, how will synthetic biology creations behave from generation to generation or over multitudes of generations? Thus, the regulation of synthetic organisms may no longer rest on a claim of substantial equivalence to their well-known unmodified counterparts. Third, will the complex genetic sequences continue to function as they did in their original sources? Or, will the reaction among the various components lead to different functions, behaviors, or unexpected emergent properties, including the evolution of new and potentially harmful characteristics, not seen in the original sources? The mutations in the genome of the synthetic organisms could result in an unexpected interaction with other living, natural organisms. A synthetic organism's possible advantages over natural species may lead to an unexpected proliferation of the synthetic biology creature, radically changing the ecosystems in unforeseen ways.

In assessing these questions, two different synthetic biology biosafety risks must be distinguished.[10] One risk involves the accidental release of a synthetic microorganism from a contained environment, such as a laboratory or a commercial production facility. If the synthetic microorganism is infectious, pathogenic, or toxic it could pose a risk to on-site workers, the health of nearby communities, and the environment upon its escape from a contained area. It is uncertain whether the synthetic organism will be able to survive in the natural environment, reproduce, and spread.

A second risk involves the potential public health and environmental risks posed by a synthetic life form intentionally designed for a noncontained use. For example, a synthetic microorganism used for commercial chemical production may be engineered to survive and function in the environment where it is released. This organism designed for a specific task could interact with naturally occurring species and cause unexpected side effects, adversely impact the environment, for instance, by affecting existing species, perhaps passing synthetic genes to natural organisms and contaminating their gene pool. The synthetic organism might evolve with new and potentially harmful characteristics creating additional uncertainties.

Federal regulators in the United States have little experience with the potential risks posed by the possible impact and/or evolution of genetically engineered microorganisms intended for use outside the laboratory. As a genetically engineered organism departs more and more from its known donor organism's genetic sequence, it becomes more difficult for risk assessors to predict the impact and/or evolution of the engineered species. Thus, the complexity made possible by synthetic biology creates uncertainty for conducting risk assessments needed to predict outcomes and design to appropriate controls. It is uncertain whether the existing risk assessment methods used by researchers will be adequate for predicting more complex changes produced by synthetic biotechnology which involves engineering entire biochemical pathways. A fear exists that errors by scientists could produce organisms that might run amok, unleashing predatory bugs with devastating public health and environmental consequences.

Researchers have adopted measures to tackle these fears. Over the years, with rDNA experiments and commercial endeavors, to decrease (if not eliminate) potentially harmful capabilities, reducing the viability of genetically engineered bacteria outside laboratories or other facilities became commonplace. For example, organisms were made dependent on nutrients that did not readily occur in a natural environment. Today, researchers engineer synthetic biology creatures so that these organisms cannot replicate themselves outside of controlled environments. Thus, as a result of self-regulation, having genomes extensively modified by synthetic biotechnologies, as opposed to a few genes taken from another organism, may not need to be subject to a greater degree of scrutiny by U.S. regulators than current measures for rDNA products and processes.

Federal Regulatory Actions

To deal with the various risks posed by synthetic biotechnologies, in 2004, a report by the National Research Council (NRC), a private, nonprofit group of distinguished scholars engaged in scientific and engineering research, recommended, among other proposals, the creation of a National Science Advisory Board for Biodefense to advise the federal government on strategies for minimizing the potential misuse of information and technologies from life sciences experimentation, considering both national security concerns and the needs of the research community.[11] In response, the federal government created the National Science Advisory Board for Biosecurity (NSABB), an advisory committee to the Secretary of the U.S. Department of Health and Human Services, the Director of the NIH, and all federal agencies that conduct or support life sciences research, to provide advice to federal departments and agencies on ways to minimize the possibility that the knowledge and technologies emanating from biological research could be used to threaten public health or national security. In December 2006, the NSABB released a report on biosecurity concerns related to the synthesis of select agents, those regulated biological agents and toxins having the potential to pose a severe threat to public, animal, or plant health or to animal or plant products.[12] The NSABB recommended that the language and implementation of existing biosafety guidelines be examined to ensure that they provide adequate guidance for those working with synthetically derived nucleic acids.[13] With the advice of the NIH Recombinant Advisory Committee, the group responsible for advising the NIH

Director on all aspects of rDNA technology, the NIH proposed revising the existing NIH Guidelines.

In March 2009, the NIH proposed modifying its rDNA guidelines to specifically cover research with nucleic acid molecules made solely by synthetic means and provide clearer guidance to scientists on how to manage synthetic biology research by setting forth principles and procedures for the risk assessment and management for such experimentation.[14] To cover synthetic genetic constructions that could pose safety risks, the proposed amendments encompass nucleic acids synthesized chemically (or by other means) apart from the use of rDNA techniques. The proposals continue the existing pattern in the NIH Guidelines of requiring a containment level commensurate to the risk of the research to the laboratory workers, the public, and the environment, with increasing containment levels as an organism (an agent, in the Guidelines' terminology) becomes more seriously pathogenic to humans. High-risk studies would continue to be subject to review by the NIH and the RAC. Factors considered in determining the containment level include virulence, pathogenicity, infectious dose, environmental stability, rate of speed communicability, quantity, availability of vaccine or treatment, and physiological activity.

It is difficult, however, to characterize the risk of an organism with an extensively modified genome or one assembled from scratch or from multiple sources. To deal with these uncertainties, the NIH proposal recommended that it would be prudent to begin by considering the highest risk group of any source of genetic material. The initial assumption is that such sequences will have the same function as they had in the original host, recognizing that a synergistic function could lead to a higher risk profile (and thus a higher containment level) than that of the contributing organism or sequence. However, making a microorganism subject to the highest level of biosafety containment requirements, in the absence of pinpointing its risk, would seem to impose unnecessary, significant costs, including impeding research and the development of potentially beneficial products and processes.

Besides the biosafety uncertainties posed by synthetic biology, another problem exists. The NIH lacks resources to ensure its institutional grantees comply with its guidelines. The NIH relies on the self-reporting of problems or violations at its funded-institutions. To date, this approach appears to be successful, again relying on the trustworthiness of the scientific community.

Another Proposed Regulatory Framework

To deal with the oversight issue, among other concerns, an expert faculty member from MIT, together with several experts from the Venter Institute and the Center for Strategic and International Studies, funded by a grant from the Alfred P. Sloan Foundation, teamed up to examine the biosafety and biosecurity issues. Their 2007 report (the Venter–Sloan Report) presented various options, but did not make any specific recommendations which option (or options) to pursue. One option, less restrictive than a ban on certain experiments, focused on continued reliance on the existing IBC oversight mechanism[15] that was established in the United States to assess the biosafety and environmental risks of rDNA experiments. Other options considered in the Venter–Sloan Report included broadening IBC responsibilities to consider the biosafety risks associated with dangerous experiments, possibly coupled with oversight from some type of national advisory group to evaluate risky research and/or enhanced enforcement of compliance with biosafety rules and guidelines.[16]

Critics of self-regulation by the science community attacked reliance on IBCs in the biosafety area. According to Edward Hammond, Director of the Sunshine Project, a biotechnology and bioweapons watchdog, "Institutional Biosafety Committees are a documented disaster. IBCs aren't up to the existing task of overseeing genetic engineering [rDNA] research, much less ready to absorb new synthetic biology and security mandates."[17]

Viewing self-regulation via IBCs more positively, at present, it seems reasonable to continue working within the existing IBC framework, mandating additional members with a broad array of relevant expertise. As restructured, IBCs will then possess the knowledge and experience to fulfill their expanded role, rather implementing some type of democratic oversight of synthetic biology as some have vaguely recommended.[18]

Besides self-regulation, a need exists to educate synthetic biology researchers on safety issues and procedures. The Venter–Sloan Report explored various educational options dealing with risks of, and guidance on best practices for synthetic genomic experiments at higher education institutions, the production of a safety manual specifically tailored for synthetic biology laboratories, and a clearinghouse for best practices.[19]

Bioterrorism and National Security Aspects

Synthetic biology research and commercialization raises concerns about biosecurity. Using biotechnologies, terrorists could, for example, assemble a superbug, a virulent pathogen. Much scientific knowledge that would be useful to malevolent individuals already exists in the public domain. Anyone with a computer can access publicly available DNA sequence databases via the Internet, obtain free DNA design software, and place an order with a commercial gene synthesis company for synthesized DNA. Even today, some DNA sequence firms do not check customers and screen DNA orders.

The creation of genomes in the laboratory has demonstrated the potential of synthetic biology to engineer harmful pathogens. For example, in 2002, a team of researchers assembled the poliovirus from off-the-shelf DNA nucleic acids in the laboratory, starting with purified oligonucleotides and instructions for the poliovirus genomic sequence.[20] Then, in 2005, scientists reconstructed the genome of the deadly 1,918 strain of influenza virus, using samples taken from the frozen cells of victims to generate a genetic sequence to copy.[21] These two examples show the potential of synthetic biology, in rogue hands, to engineer malevolent pathogens. With improved DNA synthesis technology, the assembly of larger viruses has become feasible. In malevolent hands, the possibility of combining accessible genomic data and DNA-synthesizing capabilities could be used to engineer the genomes of deadly pathogens, especially by recreating known pathogenic viruses in the laboratory and making them even more virulent.

Stepping back from these fearful scenarios, the Royal Academy of Engineers in the United Kingdom concluded that synthetic biology opponents overstated the bioterrorism risk from DNA synthesis, at least for the foreseeable future.[22] Engaging in high-tech DNA synthesis would not, at present, be a cost-effective strategy for most terrorist groups or lone operators. To create a deadly, artificial pathogen, a terrorist would need to assemble several (or more) genes that working together would enable a microbe to infect humans and cause severe illness and death. Designing such an organism to be contagious and spread from one human being to another, presents an even more difficult level of complexity. In short, it is not yet that simple to create a pathogenic organism and release it in an effective manner. Moreover, bioweapons today lack the predictable and dramatic effects of traditional modes of terrorism, such as explosives. At present, high-tech bioterrorism poses a lesser threat than lower tech forms of bioterrorism, such as anthrax, that require much less effort and are easier to produce and use.

Looking to the future, however, further advances may make DNA synthesis, together with the public availability of DNA sequence data and explanations of synthetic biotechnologies, more attractive to terrorists. Thus, a need exists for an ongoing review of the bioterrorist threat posed by synthetic biology tools.

Regulation of the Synthetic Biology Biosecurity Threat

Numerous nonprofit, governmental, and various other groups in the United States have proposed measures to deal with the biosecurity threat posed by synthetic biology. The NRC recommended the extension of biosafety measures to cover "experiments of concern" in addition to specified pathogens.[23] The NRC proposed that the U.S. Department of Health and Human Services create a review system for seven class of experiments involving microbial agents that raise concerns about their potential for misuse. These experiments include: rendering a vaccine ineffective; conferring resistance to therapeutically useful antibiotics or antiviral agents; enhancing the virulence of a pathogen or rendering a nonpathogen virulent; increasing transmissibility of a pathogen; altering the host range of a pathogen; enabling the evasion of diagnostic detection modalities; and enabling the weaponization of a biological agent or toxin. As noted earlier in this chapter, the NRC also recommended the creation of a National Science Advisory Board for Biodefense.

In 2007, the newly formed NSABB (Board) proposed a framework for the oversight of dual use, i.e., constructive and malevolent, life science research.[24] The NSABB recommended that a principal investigator conduct an initial evaluation of his or her research for its potential as dual use research of concern. The Board developed guidelines for risk assessment and management of research information.[25] In addition to proposing that research information with dual use potential be communicated responsibly,[26] the NSABB recommended the development of a code of conduct for scientists and laboratory personnel for adoption by professional organizations and institutions engaged in the performance of life sciences research.[27] To date, no code of conduct outlining scientists' responsibilities for preventing misuse of their research has been adopted.

Similar to the Asilomar conference three decades earlier, scientists entered into the dialogue regarding the biosecurity implications of synthetic biology research. In June 2004, the first international conference, Synthetic Biology 1.0, devoted to the new field was held at the Massachusetts Institute of Technology. In addition to technical presentations by researchers, policy analysts addressed the safety, security, and ethical issues associated with synthetic biology.

A second international conference, Synthetic Biology 2.0, was held at the U.C. Berkeley, in May 2006. Organizers of Synthetic Biology 2.0 sought to pass a resolution for a self-governance proposal, in response to concerns about physical harms, designed to describe some principles for advancing this new field in a safe and effective way, while addressing the challenges to biological security and justice. The resolution grew out of the Berkeley SynBio Policy Group, a joint project of Jay Keasling's laboratory and the University of California, Berkeley Goldman School of Public Policy. After interviewing scientists working in synthetic biology field and holding a series of town meetings, the group concluded that self-regulation by the science and business communities ought to be the main policy response to the biosecurity threat. The resolution tabled at Synthetic Biology 2.0 proposed screening synthetic DNA orders likely to produce known pathogens for the presence of DNA sequences from "select agents" of bioterrorism concern.[28] Under the proposed resolution recommended that all commercial DNA synthesis firms adopt current best practice screening procedures, specifically checking synthetic DNA orders encoded for dangerous sequences, such as human pathogens, and create new industry watch lists to improve industry screening programs. Furthermore, all researchers contemplating an experiment of concern that might lead to improved bioweapons obtain independent expert advice before proceeding. The resolution was not revived at subsequent international synthetic biology gatherings.

The final text that emerged from the Synthetic Biology 2.0 conference put self-governance among possible governance approaches. In placing a text on a website for further comments, the declaration, in part, stated "...[W]e support ongoing and future discussions with all stakeholders for the purpose of developing and analyzing inclusive governance options, including self-governance, that can be considered by policymakers and others such that the development and application of biological technology remains overwhelmingly constructive."[29]

In 2007, a recycled oversight proposal again focused on biosecurity concerns.[30] The proposal presented a framework where an industry body, the International Consortium for Polynucleotide Synthesis, formed by commercial firms that make and sell synthetic DNA, would share best practices and screen software to identify synthetic DNA of interest to bioterrorists. The proposal recommended that all buyers of synthesized DNA disclose their name, home institution, and offer any biosafety information relevant to the sequences they are ordering. DNA synthesis companies would use software tools to check purchaser orders against a list of potentially dangerous sequences, a set of select agents or select sequences, to help ensure compliance with the self-governance mechanism and, if needed, flag orders aimed at synthesizing possible pathogenic genes or organisms for future review.

In presenting governance options aimed at biosecurity, the 2007 Venter–Sloan report analyzed three intervention points: DNA synthesis firms; researchers; and oversight bodies. In the DNA synthesis process, attention centered on commercial DNA synthesis firms that can manufacture virtually any type of DNA, built-to-order. One suggested option would require these firms to use approved software for screening orders for hazardous substances. Another possibility would require purchasers of synthetic DNA from commercial firms to be verified as legitimate users by the research institutions where they work. Beyond verifying purchasers, a third option would require sellers to store information about customers and their orders.[31] In assessing the details of the various options for sellers, from my perspective, although screening orders for dual-use items is appropriate; however, legitimate researchers should not be prevented from quickly and easily accessing their needed goods.

With respect to researchers and the purchasers, the Venter–Sloan Report also presented a number of options regarding those constructing a genome from bottled chemicals. These included: registration of DNA synthesizers; licensing of DNA synthesizer owners; and possibly requiring a license to buy reagents, substances taking part in chemical process, or services.[32]

In addition, the Venter–Sloan Report presented various options for educating scientists about biosecurity issues and enhancing IBC responsibilities in this area. IBC review responsibilities could be broadened to consider the biosecurity risks associated with experiments and the dangerous knowledge thereby generated, possibly coupled with biosecurity oversight from a national advisory group.

Despite the plethora of recommendations, to date, Congress has only taken limited steps to deal with biosecurity issues. Following the anthrax dissemination and deaths in the United States in the fall of 2001, the Centers for Disease Control and Prevention added a Select Agent Program with biosecurity as a major element. In 2002, Congress passed the Public Health Security and Bioterrorism Preparedness and Response Act that required institutions and persons to notify the U.S. Department of Health and Human Services of their possession of certain specified pathogens and toxins, so-called select agents, that pose a severe threat to public health and safety and thus could be used in bioterrorism.[33] Laboratories working with select agents, or synthesizing DNA segments of such agents, are covered by this law and its regulations. However, it may be difficult to identify whether or not a synthetic organism is a select agent. Problems exist with respect to sequences that may not resemble those covered under the select agent regulations. Furthermore, Congress has not passed any legislation requiring U.S. suppliers to screen synthetic DNA orders for pathogenic DNA sequences. Because the oligonucleotide trade involves several nations, any effective regulatory regime must be transnational in scope.

Any governmental measures dealing with dual-use research must be adopted and harmonized internationally. If the federal government gets too tough in the biosecurity area it could drive research and commercialization overseas.

Intellectual Property Aspects

Beyond safety and security concerns, the intellectual property issues exist, resulting from the potential impact of synthetic biology, which occupies the intersection of biotechnology and software. Through the intellectual property rights obtained by patenting synthetic organisms and processes, particularly overly broad patents, the possibility exists for the creation of "unfair" monopolies with respect to the exploration and application of synthetic biotechnology through a monopoly platform product, similar to what Microsoft's Windows operating system became in the market for personal computer software.

A patent provides the inventor with exclusive rights for a limited time period in exchange for publicly disclosing the details of an invention so that others may use their information. A patent affords the inventor with legal protection against others making or using the invention for commercial purposes during the specified period.

In addition to encouraging innovation, the patent system hopefully pushes originality into the open, so it can produce more originality, while protecting the fruits of innovators' efforts so that firms can recoup the costs of their research investment and eventually generate a profit.

The 1980, the U.S. Supreme Court in Diamond v. Chakrabarty[34] upheld the patenting of living organisms, thereby opening the door to the patenting of biological products and processes. The case, which involved the patentability of a microbe constructed to degrade crude oil, established the grounds for the patentability of genetically altered organisms, reasoning that they were "not nature's handiwork." In holding that patents may be granted for "anything under the sun that is made by man," the Court's ruling established the basis for the patentability of a living microorganism modified through human intervention if it is "the result of human ingenuity."[35]

In 1988, the U.S. Patent and Trademark Office (PTO) granted the first patent for a multicellular living organism.[36] The Harvard Oncomouse, a non-naturally occurring nonhuman, multicellular living organism, was engineered to develop cancer at a high rate, thereby facilitating the study of the disease. In Animal Legal Defense Fund v. Quigg,[37] a federal appellate court consolidated the legal challenges raised by groups opposing animal patents and then dismissed their claims for lack of standing. The court established a case-by-case review process for PTO to use, in its discretion, for granting animal-related patents.

In brief, traditional U.S. patent requirements focus on: novelty; nonobviousness; and utility.[38] Novelty centers on whether the claim is new relative to the prior "art." It cannot be too much like a previous invention. Nonobviousness turns on whether the claim is or is not obvious to a person skilled in the art. Utility looks to an invention having a demonstrated, well-defined function and a clear beneficial purpose. The bottom line: unaltered genetic material in its natural environment that makes people human cannot be patented.[39] However, once genetic material is isolated, modified, and recombined, it can be patented, provided it meets the above criteria. Because of the unnaturalness of synthetic biology products and processes, they may gain U.S. patent protection. They are human inventions, not products of nature.

Beyond these general guidelines, it is uncertain how far-reaching will the patents granted on synthetic biology products and processes be. If inventors secure expansive patent protection in order to control as much of the technology as possible, smotheringly broad patents may lock up the field and limit synthetic biology's potential by restricting collaboration and stifling further development.

To date, U.S. patents have been granted on many products and processes involved in synthetic biology including methods for building synthetic DNA strands.[40] Because synthetic biotechnology requires massive computation capacity and computer memory to carry out the design and synthesis of DNA networks, all-encompassing patent claims could create gatekeeper monopolies in this field. To date, broad, concept-level patents exist on computer systems and software used in synthetic biology. For example, one patent on the system and method for simulating the operation of biochemical systems describes genes as circuits and claims: "A system and method for simulating the operation of biochemical networks [that] includes a computer memory used to store a set of objects, each object representing a biochemical mechanism in the biochemical network to be simulated."[41]

Perhaps the most famous synthetic biology patent application was filed in May 2007. The Venter Institute applied for patents on *Mycoplasma laboratorium*, a synthetic organism its researchers created. In a U.S. Patent Application,[42] entitled "Minimal bacterial genome," the institute claims exclusive ownership of a minimal set of essential genes for sustaining a synthetic, free-living organism that can grow and replicate using those genes. At present, the PTO has not the institute's patent application for the minimal genome. That if granted, the patent might result in Venter's firm, SGI, dominating the market for biofuels production from algae, among other applications.

Assuming it is appropriate to own and control broad synthetic biology patents for private gain, some maintain that such patents will slow research and that basic innovations leading to the creation of building blocks be accessible to scientists to build on and promote further breakthroughs. Balancing the protection of inventors with the need to promote research and access to a new technology is not easy. Under U.S. patent law, scientists cannot freely conduct basic research to facilitate further discoveries. No general exemption exists in American patent law for academic research. To avoid limitations on research and education, academic researchers can avail themselves of an experimental use exemption from patent infringement claims. If the act of an alleged academic infringer is "for amusement, to satisfy idle curiosity or for strictly philosophical inquiry" according to a federal appellate court case, then the activity qualifies for the narrow and strictly limited experimental use defense.[43]

Federal guidelines seek to provide some guidance with respect to access to new technology, so as to promote basic life science research. Existing NIH Guidelines address making its funded research available to the scientific community (or be subject to licensing) to facilitate further research efforts. The 1999 guidelines provide that recipients of NIH funds are expected to ensure that unique research resources be made available to the scientific community.[44] In addition to stating that research tools need not always be patented, if patented, they should seldom be licensed exclusively to one individual or organization. If an exclusive license is needed to promote further development, the guidelines recommend that it be limited to a specified commercial field, permitting the patent holder to use and distribute the tool for other research. In 2004, NIH made similar recommendations to its grant recipients regarding the patenting and licensing of a wide array of genetic materials and technologies.[45] However, the NIH Guidelines and best practices only apply to institutions receiving NIH funds, typically nonprofit higher educational organizations and research institutions.

In sum, it is difficult to protect intellectual property to stimulate innovation while not hindering basic research. Also, the exclusive licensing of patents may result in monopoly profits for a select few but hinder the development of synthetic biology in the long-run, for example, by slowing down advances in medical testing, diagnostics, and therapeutics.

Federal (and transnational) regulation of synthetic biology research and technology is about balance. Policymakers need to create a regulatory framework to reduce the biosafety and biosecurity risks but not so strong as to hamper innovation. At present, self-regulation by the science and business communities seems the more favorable path to follow. As we gain more knowledge of the risks, year-by-year, the regulatory regime may require greater governmental involvement.

Notes

1. Jonathan B. Tucker and Raymond A. Zilinskas, "The Promise and Perils of Synthetic Biology," *New Atlantis* 12 (Spring 2006): 25–45, at 36. In contrast, the Presidential Commission for the Study of Bioethical Issues, *New Directions: The Ethics of Synthetic Biology and Emerging Technologies* (Washington, DC, December 2010), 111–74, concluded that neither new federal laws and regulations nor new oversight bodies were currently

needed because synthetic biology poses few risks in its early development stages, but recommended that government agencies scrutinize the
technology more carefully to minimize risks and foster innovation. For a
summary of the risks and potential harms involving various applications
of synthetic biology, see Ibid., 62–3, 67–8, 70–1.

2.   ETC [Erosion, Technology and Concentration] Group, *Extreme
      Genetic Engineering: An Introduction to Synthetic Biology* (ETC Group,
      June 2007), 50.
3.   Paul Berg et al., "Potential Biohazards of Recombinant DNA Molecules,"
      Science 185, no. 4148 (July 26, 1974): 303.
4.   Paul Berg et al., "Asilomar Conference on Recombinant DNA Molecules,"
      Science 188, no. 4192 (June 6, 1975): 991–94. See also Paul Berg, "Reflec-
      tions on Asilomar 2 at Asilomar 3," *Perspectives in Biology and Medicine*
      44, no. 2 (Spring 2001): 183–85.
5.   Michele S. Garfinkel et al., *Synthetic Genomics: Options for Governance*
      (Venter Institute, Center for Strategic and International Studies, MIT,
      Department of Biological Engineering, October 2007), 17.
6.   U.S. Department of Health, Education, and Welfare (HEW), National Insti-
      tutes of Health (NIH), "Recombinant DNA Research Guidelines," *Federal
      Register* 41, no. 131 (July 7, 1976): 27901–43. Presidential Commission,
      New Directions, 145 (Recommendation 11: Fostering Responsibility and
      Accountability), recommended an evaluation of the effectiveness of cur-
      rent research oversight mechanisms.
7.   HEW, NIH, "Recombinant DNA Research; Actions under Guidelines,"
      Federal Register 45, no. 20 (January 29, 1980): 6718–49, at 6746–47.
8.   Executive Office of the President, Office of Science and Technology Policy,
      Coordinated Framework for Regulation of Biotechnology, *Federal Register*
      51, no. 123 (June 26, 1986): 23302–9.
9.   For an analysis of the regulation by the EPA of genetically modified organ-
      isms with novel DNA arrangements as new chemical substances under
      the Toxic Substances Control Act, see Michael Rodemeyer, *New Life, Old
      Bottles: Regulating First-Generation Products of Synthetic Biology*, Wood-
      row Wilson International Center for Scholars, Synbio 2, Synthetic Biology
      Project (Washington, DC: Woodrow Wilson Center for Scholars, March
      2009), 35–37, 39–40. The EPA exempts developers from notifying the
      agency regarding synthetic microorganisms, provided the R&D activities
      occur in a facility covered by the NIH Guidelines or a functional equiva-
      lent. EPA, "Exemptions for Research and Development Activities," *Code of
      Federal Regulations* 40: 725.232. The EPA can only regulate microorgan-
      isms not subject to the jurisdiction of any other federal agency, such as
      the Food and Drug Administration. Presidential Commission, *New Direc-
      tions*, 92–101 summarized various oversight provisions. The commission
      recommended that the federal government undertake a comprehensive
      review to assure that the multiple federal departments and agencies having
      oversight responsibilities provide adequate protection. Ibid., 102, 127–28
      (Recommendation 4: Coordinated Approach to Synthetic Biology).
10.   Rodemeyer, *New Life, Old Bottles*, 24–26. Presidential Commission, *New
      Directions*, 128 (Recommendation 5: Risk Assessment and Field Release

Gap Analysis), recommended a risk assessment review to ensure that regulators have adequate information, specifically with respect to field release of organisms.

11. National Research Council (NRC), *Biotechnology Research in an Age of Terrorism* (Washington, DC: National Research Council, 2004), 8–10, 118–21.

12. National Science Advisory Board for Biosecurity, *Addressing Biosecurity Concerns Related to the Synthesis of Select Agents* (U.S. Department of Health and Human Services, National Institutes of Health, December 2006), 2–15.

13. Ibid., 10–11.

14. U.S. Department of Health and Human Services (HHS), NIH, Office of Biotechnology Activities, "Recombinant DNA Research: Proposed Actions under the NIH Guidelines for Research Involving Recombinant DNA Molecules (NIH Guidelines)," *Federal Register* 74, no. 41 (March 4, 2009): 9411–21. See also HHS, NIH, Office of Biotechnology Activities, "Recombinant DNA Research: Proposed Actions under the NIH Guidelines for Research Involving Recombinant DNA Molecules (NIH Guidelines)," *Federal Register* 75, no. 77 (April 22, 2010): 21008–10. Presidential Commission, *New Directions* made a number of recommendations regarding risk assessment, the potential for harm through inadvertent release of organisms produced by synthetic biology, 122–23 (Recommendation 6: Monitoring, Containment, and Control), the need for reasonable risk assessment prior to field release of organisms or commercial products involving synthetic biotechnology, 131 (Recommendation 7: Risk Assessment Prior to Field Release), and the need to assess specific security and safety risks of synthetic biology research in institutional and noninstitutional settings, 147 (Recommendation 12: Periodic Assessment of Security and Safety Risks).

15. Garfinkel, *Synthetic Genomics*, 41–42.

16. Ibid., 42–44.

17. Quoted in ETC Group, News Release, "Syns of Omission: Civil Society Organizations Response to Report on Synthetic Biology Governance from the J. Craig Venter Institute and Alfred P. Sloan Foundation," October 17, 2007.

18. ETC Group, *Extreme Genetic Engineering*, 50.

19. Garfinkel, *Synthetic Genomics*, 39–41. Presidential Commission, *New Directions*, 134 (Recommendation 9: Ethics Education), recommended ethics education for researchers and student investigators.

20. Jeronimo Cello, Aniko V. Paul, and Eokard Wimmer, "Chemical Synthesis of Poliovirus DNA: Generation of Infectious Virus in the Absence of Natural Template," *Science* 297, no. 5583 (August 2, 2002): 1016–18. See also NRC, *Biotechnology Research*, 27–28.

21. Terrence M. Tumpey et al., "Characterization of the Reconstructed 1918 Spanish Influenza Pandemic Virus," *Science* 310, no. 5745 (October 7, 2005): 77–80.

22. The Royal Academy of Engineering, *Synthetic Biology: Scope, Applications and Implications* (London: Royal Academy of Engineering, May 2009), 43.

23. NRC, *Biotechnology Research*, 5–7, 113–16.

24. National Science Advisory Board for Biosecurity, *Proposed Framework for the Oversight of Dual Use Life Sciences Research: Strategies for Minimizing the Potential Misuse of Research Information* (National Science Advisory Board for Biosecurity, June 2007), 8–31.

25. Ibid., Appendix 4, Points to Consider in Risk Assessment and Management of Research Information that is Potentially Dual Use of Concern. In April 2010, the National Science Advisory Board for Biosecurity issued a report with specific recommendations designed to ensure biosecurity. National Science Advisory Board for Biosecurity, *Addressing Biosecurity Concerns Related to Synthetic Biology* (National Science Advisory Board for Biosecurity, April 2010), 13–4.

26. Ibid., Appendix 5, Points to Consider in Assessing the Risks and Benefits of Communicating Research Information with Dual Use Potential.

27. Ibid., Appendix 3, Considerations in Developing a Code of Conduct for Dual Use Research in the Life Sciences.

28. Stephen M. Maurer, Keith V. Lucas, and Starr Terrell, *From Understanding to Action: Community-Based Options for Improving Safety and Security in Synthetic Biology* (Berkeley: University of California, Berkeley, Richard and Rhoda Goldman School of Public Policy, April 15, 2006), 2–3, 14–16, 18–24.

29. Declaration of the Second International Meeting on Synthetic Biology, Berkeley, California, May 29, 2006, http://www.syntheticbiology.org/SB-2Declaration.html (accessed November 11, 2009).

30. Hans Bügl et al., "DNA Synthesis and Biological Security," *Nature Biotechnology* 25, no. 61 (June 2007): 627–29, at 629.

31. Garfinkel, *Synthetic Genomics*, 22–25. In October 2010, the Department of Health and Human Services issued guidance for screening orders of synthetic double-stranded DNA, "Screening Framework Guidance for Providers of Synthetic Double-Stranded DNA," *Federal Register* 75, no. 197 (October 13, 2010): 62820–32. Although compliance with the guidance is voluntary, many of its recommendations reflect federal law or regulations.

32. Garfinkel, *Synthetic Genomics*, 32–33.

33. 42 USC §262a.

34. 447 US 303 (1980).

35. The quotations in this paragraph are from ibid., 309–10.

36. U.S. Patent 4,736,866 (Issued April 12, 1988).

37. 932 F.2d 920 (Federal Circuit 1991).

38. 35 USC §§101, 102, 103(a).

39. In Association For Molecular Pathology v. United States Patent and Trademark Office, Case 09 Civ. 4515 (Southern District of New York, March 29, 2010), the court struck down seven patents on two genes linked to breast and ovarian cancer. See also Andrew Pollack, "Patent Protection, Breached," *New York Times*, November 2, 2010, B1; Andrew Pollack, "In a Policy Reversal, U.S. Says Genes Should not be Eligible for Patenting," *New York Times*, October 30, 2010, B1; Amy Dockser Marcus, "Licenses Drive Gene Debate," *Wall Street Journal*, April 15, 2010, A3; Shirley S. Wang, Nathan

Koppel, and Gautam Naik, "Gene Ruling Could Have Broad Reach," *Wall Street Journal*, March 31, 2010, A6; John Schwartz and Andrew Pollack, "Cancer Genes Cannot Be Patented, U.S. Judge Rules," *New York Times*, March 30, 2010, B1.

40. U.S. Patent 6,521,427, issued February 18, 2003.
41. U.S. Patent 5,914,891, issued June 22, 1999.
42. U.S. Patent Application 200701228263, May 31, 2007. The Venter Institute filed an international patent application at the World Intellectual Property Organization on a similar invention. World Intellectual Property Organization, WO2007/047148A1, April 27, 2007.
43. Madey v. Duke University, 307 F.3d 1351, 1362 (Federal Circuit, 2002).
44. HHS, NIH, Principles and Guidelines for Recipients of NIH Research Grants and Contracts on Obtaining and Disseminating Research Resources, *Federal Register* 64, no. 246 (December 23, 1999): 72090–96.
45. HHS, NIH, Best Practices for the Licensing of Genomic Inventions, *Federal Register* 69, no. 223 (November 19, 2004): 67747–48. Presidential Commission, *New Directions*, 122 (Recommendation 3: Innovation Through Sharing), recommended the review of current research and licensing practices regarding research results involving synthetic biology.

Index

For Product Safety Concerns and Information please contact our
EU representative GPSR@taylorandfrancis.com Taylor & Francis
Verlag GmbH, Kaufingerstraße 24, 80331 München, Germany